Mathematics of Tabletop Games

Mathematics of Tabletop Games provides a bridge between mathematics and hobby tabletop gaming. Instead of focusing on games mathematicians play, such as nim and chomp, this book starts with the tabletop games played by avid gamers and hopes to address the question: which field of mathematics concerns itself with this situation?

Readers interested in either mathematics or tabletop games will find this book an engaging way to begin exploring the other topic or the connection between the topics.

Features

- Presents an entry-level exposition of interesting mathematical concepts that are not commonly taught outside of upper-level mathematics courses
- Acts as a resource for mathematics instructors who wish to provide new examples of standard mathematical concepts
- Features material that may help game designers and developers make design decisions about game mechanisms
- Provides working Python code that can be used to solve common questions about games
- Covers a broad range of mathematical topics that could be used as survey material for undergraduates curious about mathematics.

Aaron Montgomery received his B.A. in Mathematics from Pomona College and his Ph.D. in Mathematics from the University of Wisconsin-Madison. He is currently employed by Central Washington University, where he teaches in the Mathematics Department and the Douglas Honors College. He has taught various traditional courses in mathematics, mathematics education, and computer programming. He has also taught special topics courses such as Games and Politics and The Mathematics of Recreational Boardgames. He is an omnigamer whose favorites include Arkham Horror: The Card Game, Undaunted: Stalingrad, Mage Knight, and the Soulsborne video games from FromSoftware.

AK Peters/CRC Recreational Mathematics Series

Series Editors

Robert Fathauer
Snezana Lawrence
Jun Mitani
Colm Mulcahy
Peter Winkler
Carolyn Yackel

Mathematical Conundrums
Barry R. Clarke

Lateral Solutions to Mathematical Problems
Des MacHale

Basic Gambling Mathematics
The Numbers Behind the Neon, Second Edition
Mark Bollman

Design Techniques for Origami Tessellations
Yohei Yamamoto, Jun Mitani

Mathematicians Playing Games
Jon-Lark Kim

Electronic String Art
Rhythmic Mathematics
Steve Erfle

Playing with Infinity
Turtles, Patterns, and Pictures
Hans Zantema

Parabolic Problems
60 Years of Mathematical Puzzles in Parabola
David Angell and Thomas Britz

Mathematical Puzzles
Revised Edition
Peter Winkler

Mathematics of Tabletop Games
Aaron Montgomery

For more information about this series please visit: https://www.routledge.com/AK-Peters
CRC-Recreational-Mathematics-Series/book-series/RECMATH?pd=published,forthcom
ing&pg=2&pp=12&so=pub&view=list

Mathematics of Tabletop Games

Aaron Montgomery

CRC Press

Taylor & Francis Group

Boca Raton London New York

CRC Press is an imprint of the
Taylor & Francis Group, an **informa** business

AN A K PETERS BOOK

Designed cover image: PicMy/Shutterstock

First edition published 2025
by CRC Press
2385 NW Executive Center Drive, Suite 320, Boca Raton, FL 33431

and by CRC Press
4 Park Square, Milton Park, Abingdon, Oxon, OX14 4RN

CRC Press is an imprint of Taylor & Francis Group, LLC

© 2025 Aaron Montgomery

Library of Congress Cataloging-in-Publication Data

Names: Montgomery, Aaron, author.
Title: Mathematics of tabletop games / Aaron Montgomery.
Description: First edition. | Boca Raton : AK Peters/CRC Press, 2025. |
Series: AK Peters/CRC recreational mathematics series | Includes
bibliographical references and index.
Identifiers: LCCN 2024003711 (print) | LCCN 2024003712 (ebook) | ISBN
9781032468518 (hardback) | ISBN 9781032468525 (paperback) | ISBN
9781003383529 (ebook)
Subjects: LCSH: Mathematical recreations. | Board games--Mathematics. |
Card games--Mathematics.
Classification: LCC QA95 .M576 2025 (print) | LCC QA95 (ebook) | DDC
794.01/51--dc23/eng/20240209
LC record available at https://lccn.loc.gov/2024003711
LC ebook record available at https://lccn.loc.gov/2024003712

ISBN: 978-1-032-46851-8 (hbk)
ISBN: 978-1-032-46852-5 (pbk)
ISBN: 978-1-003-38352-9 (ebk)

DOI: 10.1201/9781003383529

Typeset in Nimbus Roman font
by KnowledgeWorks Global Ltd.

Publisher's note: This book has been prepared from camera-ready copy provided by the authors.

To my father, who introduced me to abstract mathematics.

Contents

Preface

This book hopes to link mathematics and hobby tabletop gaming, answering the question: What mathematics do I think about while playing tabletop games? Thinking about these mathematical topics does not necessarily improve my play but often leads to interesting mathematics. Based on the questions and discussions in online tabletop game communities, these types of musing are also interesting to others. The goal is to collect several of these topics together in one place and lay some groundwork for further exploration.

I would like to say this book is for everyone, but that would not be truthful. There are people for whom neither mathematics nor tabletop games have any appeal, and those who find one of those subjects attractive may find the other of little interest. That said, if you are interested in one of these topics and not averse to the other, this book may be a starting point in exploring the connection between the two. For those familiar with either topic, it will be apparent that the contents merely skim the surface of much deeper waters. Rather than viewing this as a textbook or a reference that will answer particular questions about tabletop games, this book hopes to answer the broader question: Can this question be answered with math? Or: Which field of mathematics would provide the tools to answer this type of question? The references provided in Appendix A can be used to learn the mathematical background or discover the various games available.

THE TABLETOP GAMES

Many of us grew up with family games or classical games like *Candy Land*, *Checkers*, *Chess*, *Chutes and Ladders*, *Clue*, *Dominoes*, *Go*, *Monopoly*, *Risk*, *Tic Tac Toe*, and *Yahtzee*. I assume readers are (or can quickly become) acquainted with their rules. These are not the games under consideration in this book. Here, the focus is on what I will refer to as hobby games. While some of these games like *Catan* (p.34)[1], *Ticket To Ride* (p.62), and *Carcassonne* (p.46) have been successful at making it into big-box stores, most are only found in specialty shops or crowd-funding sites such as Kickstarter. A history of European-style hobby games (one branch of hobby games) can be found in *Eurogames: The Design, Culture and Play of Modern Board Games* [56], and more recent discussions of these games can be found in popular media ("The rise and rise of tabletop gaming" [35] and "The Invasion of the German

---

[1] A page reference after a game title indicates the page where the game's rules are introduced.

Board Games" [36]) and academic proceedings ("Major Developments in the Evolution of Tabletop Game Design" [45]). The hobby games included here are the games being played at international game conventions (for example, Spiel Essen, GenCon, Origins, and PAX Unplugged) and smaller regional and local conventions (SHUX, OrcaCon, and Dragonfly are examples from the Pacific Northwest). A complete list of all games referenced in this book can be found in the Referenced Games section at the end of the book.

While the rules for some hobby games can be fit on a single sheet of paper, many have more complex rulesets. Some of the larger rulebooks in my collection include *High Frontier 4 All* (at 54 pages) and *Empire of the Sun* (at 56 pages). As a result, many of these games are not fully explained, and you are encouraged to explore them with the resources found in Appendix A.

THE MATHEMATICS

In most places, this book focuses on finite collections and methods (topics in a discrete math or finite math course). The most notable exception is that calculus will determine the optimal bidding strategy in the chapter on auctions. The accessibility of this content to an interested undergraduate is evidenced by the wealth of undergraduate research projects involving hobby games that can be found at conferences and online. I have taught undergraduates all of the material included here. Some of the material (in particular, the content in probability and game theory) has been taught to a broad audience in general education courses. Combinatorics, graph theory, and logic have been taught in mathematics and computer science courses. Geometry, group theory, and number theory have been taught in mathematics and mathematics education courses.

Acknowledgment

I would like to thank everyone who helped develop this book, starting with my wife, Michelle, who has supported me through this process from my early mathematics courses in college to the conclusion of this project (and beyond), as well as the other members of my family who have offered encouragement in the process: Dick, Salli, Thatcher, Jaelle, and Conor.

My proofreaders deserve credit for providing quick feedback as I completed the manuscript. Jean-Marie, Jim, Kathy, Michelle, and Stuart all provided valuable feedback, from suggested reworking of large sections or even just responding to the initial garbled word stew I served up with the phrase "Huh?" Particular kudos to Michelle, who worked her way through the text even without an interest in mathematics, and Kathy, who has patiently tried to explain basic probability to me.

I also want to acknowledge my frequent gamemates (Andy, Conor, Jaelle, Martin, Meri, Michelle, Stuart, and Thatcher). Without their willingness to play games, there would not have been enough source material for this book.

Finally, I would like to thank Callum Fraser, Mansi Kabra, and all the other people at CRC Press, whose names I do not know, who helped with the production. Their help with the production was invaluable.

About the Author

Aaron Montgomery received his B.A. in Mathematics from Pomona College and his Ph.D. in Mathematics from the University of Wisconsin-Madison. He is currently employed by Central Washington University, where he teaches in the Mathematics Department and the Douglas Honors College. He has taught various traditional courses in mathematics, mathematics education, and computer programming. He has also taught special topics courses such as *Games and Politics* and *The Mathematics of Recreational Boardgames*. He is an omnigamer whose favorites include *Arkham Horror: The Card Game*, *Undaunted: Stalingrad*, *Mage Knight*, and the *Soulsborne* video games from FromSoftware.

Image Credits

The author produced all diagrams in TikZ except for Figures 2.23c and 2.24b, which are public domain. The author photographed all images from games with permission from the copyright owners. All photographed games were either owned by him or his gamemates. Images of game components from the following games are included in this book. The game components may be copyrighted, trademarked, and/or registered to the game developers and/or publishers, who retain all rights.

p. 1 *Brass: Birmingham*. Brass: Lancashire and Brass: Birmingham images and text were used with permission from Roxley Games. They are copyrighted by Roxley Games, all rights reserved worldwide.

p. 7 *Cascadia*. Cascadia illustrations, artwork, graphic design and trade dress, and associated trademarks are copyrights and/or trademarks owned by Alderac Entertainment Group Inc. or Flatout Games LLC and are used with permission.

p. 24 *Undaunted: Normandy*. Undaunted: Normandy by David Thompson & Trevor Benjamin ©2021 Osprey Publishing Ltd.

p. 32 *Hoplomachus: Remastered*. Hoplomachus: Remastered and Hoplomachus Victory images and text were used with permission from Chip Theory Games. They are copyrighted by Chip Theory Games, and all rights are reserved worldwide.

p. 34 *Watergate*. Watergate images and text were used with permission from Capstone Games. They are ©2019 Capstone Games, ©2019 Frosted Games. All rights are reserved worldwide.

p. 34 *Catan*. Catan images and text were used with permission and are copyright ©1995 CATAN GmbH - CATAN, the CATAN logo, the "CATAN Sun," and the CATAN Brand Logo are trademark properties of CATAN GmbH (catan.com). All rights reserved.

p. 38 *My City*. "My City" by Reiner Knizia, Michael Menzel ©2020 Franckh-Kosmos Verlags GmbH & Co. KG, Stuttgart, Germany. "My City" by Reiner Knizia, Michael Menzel ©2020 Thames & Kosmos, LLC, Providence, RI, USA

p. 38 *The Isle of Cats*. Images and photographs of The Isle of Cats are copyrighted to The City of Games Limited.

p. 39 *Hamlet: The Village Building Game*. Hamlet images used with permission from owners: Mighty Boards/David Chircop.

CHAPTER 1

Combinatorics

Figure 1.1: Components from *Brass: Birmingham*.[1]

In the tabletop game *Brass: Birmingham*, players assume the role of entrepreneurs during the Industrial Revolution. Cards determine the ability to build factories and extend the industrial network through canals and railroads. At the start of the game, each player draws eight cards from the 64-card deck, as shown in Figure 1.1. Knowing the likelihood that one's opponents have a particular card in their hand at the start of the game can affect a player's strategy. As a start to determining this likelihood, this chapter will answer the related question.

How many different opening hands are possible in a game of *Brass: Birmingham*?

---

[1]Brass: Lancashire and Brass: Birmingham images and text were used with permission from Roxley Games. They are copyrighted by Roxley Games, all rights reserved worldwide.

INTRODUCTION

Counting is one of the earliest uses of mathematics and is found throughout this work, as many questions about tabletop games can be answered directly or indirectly by careful counting. While the basic idea behind counting is straightforward (pointing at each item while reciting whole numbers), this process can fail for two reasons.

First, in some situations, duplicates may cause this straightforward but naïve counting technique to overcount the number of distinct objects. More advanced techniques are developed to handle these duplications through the initial sections of this chapter. In a mathematics course, questions often take the form of "How many ways can one distribute 13 indistinguishable objects into three distinguishable boxes?" No one, mathematician or otherwise, spends their weekends placing indistinguishable objects into distinguishable boxes. However, tabletop games provide an accessible application of these counting rules.

Second, in many situations of interest, the number of objects to be counted far exceeds our ability to process all of them. Even if the fastest supercomputer (circa 2023) had been listing shuffles of a deck of standard playing cards since the Big Bang, it would still have only listed a minuscule percentage of all possible shuffles. Understanding the size of the numbers involved may help one determine a reasonable solution technique.

1.1 SUM RULES

Some more straightforward questions will be pursued before tackling the opening question about *Brass: Birmingham*. The answers to these questions will establish some of the notation and basic counting rules used throughout the text.

In the cooperative game *The Grizzled*, players take the roles of World War I soldiers attempting to survive trench warfare. They do this by playing trial cards that contain different military threats (an assault, a gas attack, or shelling) and attempting to avoid creating a set of three matching threats. As the game progresses, the characters will collect hard-knock cards that hinder their ability to perform. Three hard-knock cards are phobias, three hard-knock cards are traumas, and thirteen hard-knock cards are neither. How many total hard-knock cards are there?

The natural inclination is to add the number of each type of hard-knock card to determine that there are $3 + 3 + 13 = 19$ hard-knock cards. In this case, this technique works and is an example of the Sum Rule.

Sum Rule: If there are m objects with one trait and n objects with a second trait and no object has both traits, then there are $m + n$ objects with at least one of the two traits.

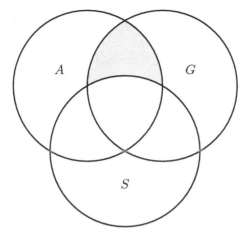

Figure 1.2: Venn Diagram of Cards in *The Grizzled*.

In the threat deck, there are 14 cards containing each of the three threats. How large is the deck of threat cards?

While using the Sum Rule here might be tempting, that rule doesn't apply since some threat cards contain more than one threat. Instead, the number of cards with multiple threats must be tracked carefully. This is best explained using a Venn diagram.[2] A Venn diagram consists of several (often circular) overlapping regions labeled by the traits of the objects under consideration. The points inside each region represent the objects with the label of that region. In this case, there are three traits: assault, labeled A; gas, labeled G; and shelling, labeled S. This produces the Venn diagram with three circles shown in Figure 1.2. Each circle is split into four subregions. The subregion with the label represents cards that contain only one threat. For example, the region labeled A is where the cards with only the assault threat would be placed. The regions shaped like the shaded region contain cards with exactly two threats. For example, the shaded region contains those cards that have both the assault threat and the gas threat but not the shelling threat. Finally, the region in the center of the diagram contains the cards with all three threats.

One could carefully separate the cards into each category for a small collection. That might even be easier in this case than the Inclusion-Exclusion Principle on page 6. Still, in many cases, starting with the Sum Rule and then modifying the final count to accommodate the overlapping regions will be easier. Starting with the Sum Rule, the sum of the assault, gas, and shelling cards is the initial estimate. A tally mark will be placed in the Venn diagram each time the cards in that region are counted toward the result. The tally marks do not represent the number of cards in the region; they represent the

---

[2]Venn diagrams are named after John Venn, an eighteenth- to nineteenth-century English mathematician who formalized their use.

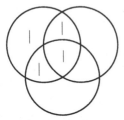

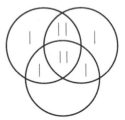

(a) Marks from the assault census taker.

(b) Marks from the assault and gas census takers.

Figure 1.3: Counting with the Inclusion-Exclusion Principle (Part 1).

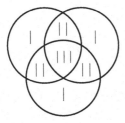

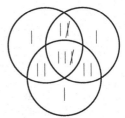

(a) Marks from all three census takers.

(b) After removing the double-count between the assault and gas census takers.

Figure 1.4: Counting with the Inclusion-Exclusion Principle (Part 2).

number of times the region has been inventoried. One way to imagine this is to consider the regions as rooms, each containing cards matching the associated traits. All cards with an assault and a gas attack but not shelling on them are in the shaded room in Figure 1.2. Three census takers—one counting assault threat cards, one counting gas threat cards, and one counting shelling threat cards—walk through the rooms containing cards with those threats. When in the room, the census taker will record the number of cards and leave a mark on the floor, represented by | in the diagrams, to remind them that they have counted that room. For example, after the assault census taker has walked through the four rooms with assault cards, the floor of the rooms is shown in Figure 1.3a since the assault census taker counts cards in the four rooms in the upper-left corner of the diagram.

After the assault census taker has walked through the rooms, the gas census taker will walk through the rooms. Again, they record the number of cards in each room containing cards with the gas threat and mark the floor to indicate that they have counted that room. The floor of the rooms after the gas census taker has walked through the rooms is shown in Figure 1.3b. Finally, the shelling census taker does the same, and the floor of the rooms after the shelling census taker has walked through the rooms is shown in Figure 1.4a.

The three census takers then combine their counts to claim there are $14 + 14 + 14 = 42$ threat cards. However, when the cards are placed in one pile and counted, only 35 cards are counted, which is 7 cards short. The census takers now notice two tally marks in three regions and three tally marks in the center region. This means that those cards have been counted twice or thrice (depending on the number of marks in the region). For example, the cards in the shaded region from Figure 1.2 were counted twice: once by the assault census taker and once by the gas census taker. The cards in the center were counted thrice: once by each census taker. At this point, possibly after some angry discussion about why no one was paying attention to the marks on the floor, the census takers attempt to use their notes to determine the correct count. To track exactly what is happening, some notation may be useful. Given a set X, the notation $|X|$ is the number of objects in X, the cardinality of X. So, $|A| = 14$ indicates that there are 14 assault threat cards. The set $X \cup Y$ is the union of X and Y and is the set of objects that are either in X or Y (objects which have at least one of the two traits). Finally, the set $X \cap Y$ is the intersection of X and Y and is the set of objects that are both in X and Y (objects which have both traits). The current estimate is $|A| + |G| + |S| = 14 + 14 + 14 = 42$. This number will be too large because some cards were counted more than once.

The overcount caused by the overlap between the assault and gas census takers can be remedied by subtracting the number of cards in the intersection of the assault and gas regions,

$$|A| + |G| + |S| - |A \cap G| = 14 + 14 + 14 - 3 = 39.$$

To represent that the three cards double-counted by being in both the assault and gas census taker's rooms were removed from the count, one tally mark in each of the regions is erased, represented by the symbol $/$ in the diagram. The result is shown in Figure 1.4b. At this point, the cards in four rooms have each been counted once, and the cards in three rooms have been counted twice.

Repeating this process for the overlap between the assault and shelling census takers and the gas and shelling census takers leads to the sum,

$$|A| + |G| + |S| - |A \cap G| - |A \cap S| - |G \cap S|$$
$$= 14 + 14 + 14 - 3 - 3 - 3 = 33.$$

This value still needs to be corrected, as it is short two cards. Returning to the marks on the floor, the census takers realize that one mark was erased from the center room each time a double-intersection was subtracted. The result is that no marks remain in the center room, as shown in Figure 1.5a, so those cards have yet to be counted in the current formula.

The correct count is obtained by adding the number of cards in that region back into the sum. So, the formula will be to add the size of each set, subtract the size of the double-intersections, and then add the size of the

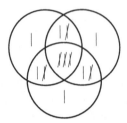

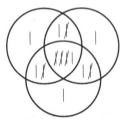

(a) After removing all double-counts.

(b) After removing all double-counts and adding back in the triple-count.

Figure 1.5: Counting with the Inclusion-Exclusion Principle (Part 3).

triple-intersections. Each of the floors of the rooms corresponding to this calculation, shown in Figure 1.5b, has exactly one tally mark, indicating that the cards in each region were counted precisely once. This formula arrives at 35 cards:

$$|A \cup G \cup S|$$
$$= (|A| + |G| + |S|) - (|A \cap G| + |A \cap S| + |G \cap S|) + (|A \cap G \cap S|)$$
$$= (14 + 14 + 14) - (3 + 3 + 3) + (2) = 35.$$

This is an example of the Inclusion-Exclusion Principle, expressed in terms of merging sets. To describe this principle, the intersections of an odd number of sets will be referred to as an "odd intersection" and intersections of an even number of sets as an "even intersection." So, a single set is an odd intersection because it is the intersection of one set. A double intersection is an even intersection because it is the intersection of two sets. A triple intersection is an odd intersection, and so forth. The Inclusion-Exclusion Principle uses this terminology to extend the Sum Rule.

Inclusion-Exclusion Principle: When counting the number of objects in the union of sets, add all the odd intersections and subtract all the even intersections.

Using formal notation, the case for two sets becomes:

$$|A \cup B| = (|A| + |B|) - (|A \cap B|)$$

While the case for three sets becomes:

$$|A \cup B \cup C| = (|A| + |B| + |C|)$$
$$- (|A \cap B| + |A \cap C| + |B \cap C|)$$
$$+ (|A \cap B \cap C|)$$

Figure 1.6: Habitat Tiles and Wildlife Tokens in *Cascadia*.[3]

A Venn diagram of the union of four or more sets does not have a nice diagram that fits on paper, but the formula continues to hold:

$$
\begin{aligned}
|A \cup B &\cup C \cup D| \\
&= (|A| + |B| + |C| + |D|) \\
&\quad - (|A \cap B| + |A \cap C| + |A \cap D| + |B \cap C| + |B \cap D| + |C \cap D|) \\
&\quad + (|A \cap B \cap C| + |A \cap B \cap D| + |A \cap C \cap D| + |B \cap C \cap D|) \\
&\quad - (|A \cap B \cap C \cap D|)
\end{aligned}
$$

1.2 PRODUCT RULES

While the Sum Rule and Inclusion-Exclusion Principle apply to situations where one object is selected from multiple sources, there are several situations where multiple selections must be made.

In *Cascadia*, players select and place habitat tiles and wildlife tokens in a tableau in front of them, as shown in Figure 1.6. Each habitat tile shows one or two habitats and which wildlife tokens can be placed on it. Each wildlife token shows one of five types of animals from the Pacific Northwest. Players will score points based on the patterns of animals on their habitat tiles.

There are 25 tiles that show exactly one habitat, and there are 60 tiles that show exactly two habitats. How many total tiles are there in *Cascadia*? Similarly, there are 20 wildlife tokens for each of the five animal types (each token only displays one animal type). How many total wildlife tokens are there?

Since no single tile can show one habitat *and* two habitats, each has exactly one of these traits. Therefore, the Sum Rule calculates that there are 25 +

---

[3] Cascadia illustrations, artwork, graphic design and trade dress, and associated trademarks are copyrights and/or trademarks owned by Alderac Entertainment Group Inc. or Flatout Games LLC and are used with permission.

$60 = 85$ tiles. Since no token has more than one animal on it, the Sum Rule calculates that there are 100 wildlife tokens:

$$20 + 20 + 20 + 20 + 20 = 5 \times 20 = 100.$$

In both situations, only a single selection is made, either selecting a single habitat tile or a single wildlife tile. At the start of the game of *Cascadia*, one of the 85 habitat tiles is randomly paired with one of the 100 wildlife tokens. How many pairs are possible?

In this case, for each of the 85 habitat tiles, there are 100 possible options for the wildlife token. Let's take the collection of all possible pairs and break it into 85 types of pairs based on which habitat tile is in the pair. Each type will have 100 pairs, as there are 100 wildlife tokens that could be paired with the tile. So, there are 85 collections of 100 pairs each, and these collections do not share any common elements (as they all have a different habitat tile). As a result, the Sum Rule calculates that there will be 8500 pairs:

$$\underbrace{100 + 100 + \cdots + 100}_{85 \text{ terms}} = 85 \times 100 = 8500.$$

However, rather than considering this as a large sum, it makes more sense to consider this as the result of making two selections: first selecting the tile and then selecting the token. When considered this way, there are 100 options for the tile and 85 options for the token. The general rule for this situation is the Product Rule.

Product Rule: If there are m objects in one collection and n objects in a second collection, then there are $m \times n$ pairs of objects where the first object comes from the first collection and the second object comes from the second collection.

Like the Sum Rule, the Product Rule can be reformulated in traditional mathematical notation, where $X \times Y$ represents pairs where the first element comes from X and the second element comes from Y, sometimes referred to as the Cartesian product of the sets. With this notation the Product Rule becomes

$$|A \times B| = |A| \times |B|.$$

Like the Sum Rule, the Product Rule result extends to more than two selections. In *Cascadia*, the players will score points based on how the wildlife tokens in front of them are placed, and each animal scores differently. The details of the scoring rules are determined by wildlife scoring cards, which are chosen randomly at the start of the game. Each of the five animal types has three different scoring cards from which one is selected. To determine a complete set of scoring rules requires selecting one of the three cards for salmon to determine how salmon tokens score, one of the three Roosevelt elk cards to determine how Roosevelt elk tokens score, and so forth. How many possible scoring combinations are there in *Cascadia*?

Here there are five selections, and with each selection, there are 3 options, so there would be 243 scoring options:

$$\underbrace{3 \times 3 \times 3 \times 3 \times 3}_{\text{5 factors}} = 3^5 = 243.$$

In this case, all the collections in the Product Rule are the same size. This specialization of the Product Rule is called the Power Rule.

Power Rule: If k selections are made, each from a collection of size n, then the total number of ways to make those selections is n^k.

The Power Rule applies when multiple items are selected from the same collection, and the selected object is not removed from the collection when it is selected (for instance, if a die is rolled multiple times). These situations will be referred to as selection with replacement.

Another common situation occurs when the selected item is removed from the collection when it is selected. This is often referred to as selection without replacement. This happens in *Cascadia* when four habitat tiles are drawn at the start of a game to pair with four wildlife tokens. How many ways can the first four tiles in *Cascadia* be selected?

While there are 85 options for the first tile selected, there are only 84 options for the second tile (since the first tile has been removed from the collection). Similarly, there are 83 options for the third tile and 82 options for the fourth tile. As a result, there are 48 594 840 ways these four tiles can be selected:

$$85 \times 84 \times 83 \times 82 = 48\,594\,840.$$

This situation is common enough that it is included as one of the fundamental counting rules and given a notation in the Permutation Rule. The name comes from the fact that in mathematics, a permutation refers to an ordering of a collection of objects.

Permutation Rule: If k selections are made from a collection of size n without replacement where the order of selection matters, then there are $P(n, k)$ ways to make those selections, where

$$P(n, k) = \underbrace{(n)(n - 1)(n - 2) \cdots (n - k + 1)}_{k \text{ factors}}$$

There is a special notation for the case, where n objects are selected without replacement from a set of n objects, the factorial. This value can be written as a permutation for $n > 0$.

$$n! = P(n,n) = \underbrace{(n)(n-1)(n-2)\cdots(1)}_{n \text{ factors}}$$

This provides an alternative definition for permutations that is more efficient than writing out each factor, especially when k is large.

$$P(n,k) = \frac{n!}{(n-k)!}.$$

While this is a good formulation for small values of n and k or for algebraic manipulation, it is not always a good choice when computing the value given large values of n and k. The issue is that the calculation is obtained by taking two products and then dividing them. Because this procedure can take longer and lead to round-off errors, many programming languages and computer algebra systems have a special function to handle permutations.

Shuffling stacks of tiles or decks of cards is one place where factorials occur. For example, there are $85! > 10^{128}$ ways to shuffle *Cascadia* tiles. Another good example of selection without replacement occurs when dealing cards from a deck. In most games, the order in which the cards are dealt is irrelevant, so a different counting technique must be used (see Section 1.3). However, in some games, players must keep the cards in the order they were dealt.

In *Bohnanza*, players take the role of bean farmers. Cards in the game show different varieties of beans, and players will play their cards from their hands to plant that bean variety in their field. Throughout the game, cards drawn must be placed in order at the back of the player's hand, and played cards must come from the front. A significant portion of the strategy in the game involves trading cards, which allows players to remove cards from the center of their hand. A similar mechanism is found in *SCOUT*, where each player portrays the leader of a circus attempting to put together successively more exciting acts. In the five-player game of *SCOUT*, each player is dealt nine cards from the deck of 45 unique cards and cannot change the order of the cards. A player will try to play a group of sequential cards in their hand stronger than the current combo (see page 82). The Permutation Rule can be used to determine that there are approximately 3×10^{14} opening hands:

$$P(45,9) = \underbrace{(45)(44)(43)(42)(41)(40)(39)(38)(37)}_{\text{nine factors}} = 321\,570\,878\,428\,800.$$

While developing the Power Rule and Permutation Rule, some assumptions were made about the values of k and n. While most situations will fit within those assumptions, extending the formulas to other values will make using the formulas more straightforward.

The first convention is that empty products are equal to one. The convention about empty products results in the following formulas:

$$0^0 = 1$$
$$0! = 1$$

In calculus, the expression 0^0 is considered indeterminate, meaning it is not assigned a value. This is because the value of $\lim_{(x,y)\to(0,0)} x^y$ depends on the relation between x and y. However, in the context of counting, the value of 0^0 is set to one since it follows the empty product rule and correctly extends many counting formulas. A consequence of the rule that $0! = 1$ is that $P(n,0) = 1$, which makes some sense in that there is only one way to select nothing from a collection of n elements (namely, to select nothing).

The second convention involves permutations.

$$P(n,k) = 0 \text{ if } k < 0$$
$$P(n,k) = 0 \text{ if } k > n$$

Both of these make intuitive sense as one cannot select a negative number of objects from a set of n objects, nor can one select more than n objects without replacement from a collection of n objects. However, they deviate from the algebraic formula that computes $P(n,k)$ when $0 \le k \le n$.

1.3 BINOMIALS AND MULTINOMIALS

While the Sum Rule and Product Rule provide the foundation of counting, they overcount when different outcomes should be considered equivalent. For example, when dealt a hand of cards in most card games, a player can rearrange the cards, so the order in which the cards are dealt is irrelevant.

Binomial Coefficients

In *Ark Nova*, players attempt to build a modern zoo and support conservation projects by playing zoo cards. At the start of the game, a player will draw eight of the 212 zoo cards and two of the eleven final scoring cards. As in most card games, the order in which a player draws these cards is irrelevant. How many possible opening hands are there in *Ark Nova*?

The final scoring cards will be calculated first, as fewer of these cards exist. There are $P(11,2)$ ways in which the two cards could be dealt to the player. However, the player only cares about which cards were drawn rather than their order. So, there is an overcount for each of the different ways a pair of cards could be drawn. For example, two cards that could have been drawn are Climbing Park and Conservation Zoo. Since there are two ways these two cards could have been dealt to the player (depending on which card arrived first), this opening hand has been counted two times in the value $P(11,2)$.

Therefore, the two final scoring cards can be selected in 55 ways:

$$\frac{P(11, 2)}{2} = 55. \tag{†}$$

For the eight Zoo cards, there are $P(212, 8)$ ways in which the cards could have been drawn. However, there are $8! = P(8, 8)$ ways in which they could have been ordered in the draw, and each of these $8!$ ways will lead to the same opening hand. So the eight Zoo cards can be selected in approximately 9×10^{13} ways:

$$\frac{P(212, 8)}{8!} = 88\,535\,640\,906\,570$$

Note that thinking about the situation for eight cards reveals that the denominator of 2 in Equation (†) is the value of $2!$, as there are $2!$ ways to order two objects.

The Product Rule provides the final answer. There are over 4×10^{15} opening hands:

$$\frac{P(11, 2)}{2!} \times \frac{P(212, 8)}{8!} = 4\,869\,460\,249\,861\,350$$

This technique is generalized in the Combination Rule.

Combination Rule: If k selections are made from a collection of size n without replacement and the order of selection does not matter, then there are $C(n, k)$, sometimes written as $\binom{n}{k}$, ways to make those selections, where

$$\binom{n}{k} = \frac{P(n, k)}{k!} = \frac{n!}{k!(n-k)!}.$$

The value $C(n, k)$ is called the binomial coefficient, and there are multiple common notations for it. In places where vertical space is constrained (for instance, when included in a paragraph), the $C(n, k)$ notation will be used. However, in places where horizontal space is constrained (for instance, in displayed equations), the $\binom{n}{k}$ notation will be used. In both cases, the expression is typically read as "n choose k." When $k < 0$ or $k > n$, $C(n, k) = 0$ following the same convention as $P(n, k)$.

The value of $C(n, k)$ was calculated above by considering how many ordered selections were associated with a single unordered selection and dividing. An alternative way to arrive at the same result would be to derive the number of ordered selections from the number of unordered selections. This is done by considering the expression $C(n, k)$ as an unknown value and then solving for it in an equation that relates it to known values. There are $P(212, 8)$ ways to order eight of the 212 Zoo cards in *Ark Nova*. One can calculate this number by first selecting which eight cards are to be ordered (there are $C(212, 8)$ possibilities) and then determining the order of those eight cards (there are 8! possibilities). The total number of orderings is the product of these two values, so $P(212, 8) = C(212, 8)8!$. In general, this reasoning gives the formula

$$P(n, k) = k! \binom{n}{k}.$$

Solving for $C(n, k)$ leads to the same value as before. Yet another way to derive this formula is presented in Section 3.3.

When computing the number of ways to shuffle a stack of *Cascadia* (p.7) tiles or a deck of *SCOUT* (p.10) cards in Section 1.2, it was noted that each tile or card differed from the others. The counting becomes more complicated if the objects can be the same, as in games like *Bohnanza* (p.10) or *Flamme Rouge*.

In *Flamme Rouge*, players take on the role of a team of two cyclists. Each cyclist has a deck of cards listing numbers between two and nine. On each turn, a player will draw three cards from each cyclist's deck and select one to play, moving their cyclist forward that many spaces. The played card will be removed from the game, while the two unplayed cards will eventually be returned to the cyclist's deck. While it may seem like a good idea always to choose the card with the highest value, the game introduces terrain, exhaustion, and drafting rules, making it advantageous to choose smaller numbers in some cases. One of the cyclists is designated as a sprinter, and their deck contains 3 cards of each value from 2 through 5 and 3 cards with value 9, for a total of 15 cards. In this game, all cards with the same value play the same. So, it matters whether there is a value-five card on the top of the deck, but not which value-five card is on the top. How many ways can a *Flamme Rouge* deck be shuffled if only the value of each card is significant?

The Permutation Rule calculates how the individual cards can be ordered. However, many of these permutations will lead to the same values occurring in the same location in the deck. For example, the value-two cards can be freely rearranged amongst themselves without changing the shuffle result, as they are indistinguishable. Since there are three of them, there are 3! duplicates from the possible permutations of the value-two cards. This same reasoning is repeated for the value-three, value-four, value-five, and value-nine cards. This results in approximately 1.7×10^8 distinct shuffles:

$$\frac{15!}{3!3!3!3!3!} = 168\,168\,000.$$

When some objects are indistinguishable from each other for game purposes, they are referred to as being only distinguishable by type. Otherwise, they are individually distinguishable. For example, the only significant feature of a card in the sprinter's deck in *Flamme Rouge* is its value, so all cards with the same value are considered to be of the same type, and the cards are distinguishable only by this type. In this situation, the phrase "only distinguishable by value" is used to specify that the card's value determines its type. This situation is handled by the Multinomial Rule. Given the generality of this situation, the notation will be complicated as it must handle all the possible cases.

Multinomial Rule: If there are $n = k_1 + \cdots + k_t$ objects of t types where there are k_i objects of the ith type and the objects are only distinguishable by type, then there are $M(k_1, \ldots, k_t)$, sometimes written $\binom{n}{k_1, \ldots, k_t}$, ways to order the objects, where

$$\binom{n}{k_1, \ldots, k_t} = \frac{n!}{k_1! \ldots k_t!}.$$

The value $M(k_1, \ldots, k_t)$ is called the multinomial coefficient, and there are multiple common notations for it. The notation $M(k_1, \ldots, k_t)$ will be used when vertical space is constrained, and the notation $\binom{n}{k_1, \ldots, k_t}$ will be used when horizontal space is constrained. Like the binomial coefficients, the definition is expanded to situations, where the formula is no longer valid by setting $M(k_1, \ldots, k_t) = 0$ if any of the $k_i < 0$.

The multinomial coefficients generalize the binomial coefficients. To select k objects from a collection of n objects, shuffle a deck of k green cards and $n - k$ red cards, then assign cards to the objects. Objects assigned a green card are included in the selection, and objects assigned a red card are not. As a result, each selection corresponds to a shuffling of the card deck:

$$\binom{n}{k} = C(n, k) = M(k, n - k) = \binom{n}{k, n - k}.$$

Note that while the two expressions look similar, the expression on the far left is a binomial coefficient (with only one lower index), but the expression on the far right is a multinomial coefficient (with two lower indices). The extended cases in the binomial coefficient, that $C(n, k) = 0$ when $k < 0$ or $k > n$, match the cases in the multinomial coefficient, where $M(k, n - k) = 0$ when $k < 0$ or $n - k < 0$, so the two values match in these cases as well.

The permutations are also a special case of the multinomial coefficients. For a permutation of k objects from a collection of n objects, shuffle a deck of k green cards that are numbered from 1 to k and a collection of $n-k$ red cards. There are now k cards individually distinguishable by the number printed on them and $n - k$ cards indistinguishable from each other. Again, distribute the cards, but now the numbers printed on the green cards are used to order the selected objects:

$$P(n, k) = \binom{n}{\underbrace{1, \ldots, 1}_{k \text{ 1s}}, n - k}.$$

The Multinomial Rule can be explained in two alternative ways using previous counting techniques (and a third will be presented in Section 3.3). These explanations will be presented using the specific example of the *Flamme Rouge* deck. The first matches the alternative explanation for the binomial coefficient. There are 15! ways to order the entire deck of cards. One way to calculate this value is first to determine the positions associated with each card value (there are $M(3, 3, 3, 3, 3)$ possibilities) and then decide which of the value-two cards are placed in the three positions assigned a value-two card (there are 3! possibilities), and so forth. This arrives at $15! = M(3, 3, 3, 3, 3)3!3!3!3!3!$. The general formula would be

$$n! = k_1!k_2! \cdots k_t! \binom{n}{k_1, \ldots, k_t}.$$

As expected, this is consistent with the formula in the Product Rule.

For the second alternative, first, decide where to place the three value-two cards in the 15 positions of the deck ($C(15, 3)$ possibilities). Once this is done, determine where to place the three value-three cards in the remaining twelve positions in the deck ($C(12, 3)$ possibilities). Continue this until the three value-nine cards are placed in the remaining three positions. Multiplying all of these values together,

$$\binom{15}{3, 3, 3, 3, 3} = \binom{15}{3}\binom{12}{3}\binom{9}{3}\binom{6}{3}\binom{3}{3}.$$

To see that this is the same value as presented in the Multinomial Rule, expand each binomial coefficient using the formula in the Combination Rule and use the fact that $0! = 1$:

$$\binom{15}{3}\binom{12}{3}\binom{9}{3}\binom{6}{3}\binom{3}{3} = \frac{15!}{3!12!} \cdot \frac{12!}{3!9!} \cdot \frac{9!}{3!6!} \cdot \frac{6!}{3!3!} \cdot \frac{3!}{3!0!} = \frac{15!}{3!3!3!3!3!}.$$

The general formulation in this case is

$$\binom{n}{k_1, \ldots, k_t} = \binom{n}{k_1}\binom{n - k_1}{k_2} \cdots \binom{n - (k_1 + \cdots + k_{t-1})}{k_t}.$$

1.4 STARS, BARS, AND GENERATING FUNCTIONS

Recall that in *Flamme Rouge* (p.13), players draw 3 cards from a deck of 15 cards consisting of 3 cards each of the values 2 through 5 and nine. How many opening hands are possible?

Because the opening hand is not ordered and the cards are only distinguishable by type, the Product Rule seems like a good start. However, tracking the details soon becomes overwhelming. For example, if the opening hand contains two value-two cards, these cards could be reordered. However, not all of the value-two cards will be permuted as one of those cards remains in the deck. To solve this problem, something different is needed, such as a technique popularized by William Feller[4] known as stars and bars. The technique relies on a sequence of symbols, some of which are "stars," representing the objects, and some of which are "bars," dividing the objects into different types. The symbol $\star$ will represent the stars, and the symbol | will represent the bars.

In the *Flamme Rouge* example, there are five types of cards, as there are five values. Therefore, the sequence will be divided into five regions using four bars. Notice that there is one fewer divider than there are types. Stars in the region to the far left will represent value-two cards, stars in the next region will represent value-three cards, and so forth until the region to the far right represents value-nine cards. For example, the sequence

$$\underbrace{\star\star}_{\text{value-two cards}} \mid \underbrace{\star}_{\text{value-three cards}} \mid \underbrace{}_{\text{value-four cards}} \mid \underbrace{\star\star\star}_{\text{value-five cards}} \mid \underbrace{\star}_{\text{value-nine cards}}$$

would represent two value-two cards, one value-three card, no value-four cards, three value-five cards, and one value-nine card.

For an opening hand in *Flamme Rouge*, there are three stars (the cards) and four bars (separating the five types). The Combination Rule can compute the number of sequences by deciding which three of the seven possible locations will be the stars, leaving the remaining four positions for the bars. So there are $C(7,3) = 35$ possible opening hands. This calculation is accurate because every sequence of symbols corresponds to a possible opening hand, and every opening hand corresponds to a sequence.

Stars and Bars Technique: If there are t types of objects, objects are only distinguishable by type, and there are at least n objects of every type, then the number of ways to select n objects is given by

$$\binom{n + (t - 1)}{n}.$$

Later in the game, after some cards have been removed, not every sequence that can be constructed will correspond to a possible hand. For example, assume that it is late in the game, and the deck is down to two value-two cards, three value-three cards, one value-four card, no value-five cards, and two value-nine cards. At this point in the game, the sequence

$$\star || \star \star ||$$

would no longer be possible as it requires two value-four cards, which are no longer in the deck. How many possible hands can be drawn from the eight cards listed above?

---

[4]William "Vilim" Feller was a twentieth-century Croatian-American mathematician who popularized probability in America.

One technique for handling this would be determining which sequences are impossible and subtracting these from the count. This would work, but generating functions will allow a computer to do most of the computations. The full theory of generating functions is a very versatile counting technique and requires the mathematical technique of infinite formal sums. However, everything can be kept finite for the cases under consideration, so they only rely on polynomial multiplication.

Generating functions work by encoding information into an expression whose coefficients answer the question. A formal variable, x, is used to do this. The value of x does not represent an unknown or a changing quantity (as is typical in algebra or calculus) but is there to facilitate the calculation. The values of interest will be the coefficients of terms in a polynomial in x. A selection of n objects is encoded as x^n and a sum is used to indicate a selection from one of the terms. So a selection of zero to three objects is encoded in the expression $x^0 + x^1 + x^2 + x^3$ (where x^0 is often reduced to one). If multiple selections are made, the product of the expressions encodes the result. For example, because x^1 represents selecting one object, $x \times x$ would represent selecting that object twice (matching the use of x^2 earlier). So, selecting cards from a full *Flamme Rouge* deck can be represented as $p(x)$, where

$$p(x) = \underbrace{\left(1 + x^1 + x^2 + x^3\right)}_{\text{value-two cards}} \times \underbrace{\left(1 + x^1 + x^2 + x^3\right)}_{\text{value-three cards}}$$
$$\times \underbrace{\left(1 + x^1 + x^2 + x^3\right)}_{\text{value-four cards}} \times \underbrace{\left(1 + x^1 + x^2 + x^3\right)}_{\text{value-five cards}} \times \underbrace{\left(1 + x^1 + x^2 + x^3\right)}_{\text{value-nine cards}}.$$

Expanding $p(x)$ leads to the expression

$$p(x) = 1 + 5x^1 + 15x^2 + 35x^3 + \cdots + 5x^{14} + x^{15}.$$

This indicates that there is one way to select zero cards (the coefficient on x^0 is one), that there are five ways to select one card (the coefficient on x^1), that there are 15 ways to select two cards, and there are 35 ways to select three cards. This matches the previous calculation that there are 35 ways to select an opening hand of three cards.

The late game restrictions above modify this calculation by limiting how many cards can be chosen for each type. Instead of the expression above, the factors have more limited exponents,

$$p(x) = \underbrace{\left(1 + x^1 + x^2\right)}_{\text{value-two cards}} \times \underbrace{\left(1 + x^1 + x^2 + x^3\right)}_{\text{value-three cards}}$$
$$\times \underbrace{\left(1 + x^1\right)}_{\text{value-four cards}} \times \underbrace{(1)}_{\text{value-five cards}} \times \underbrace{\left(1 + x^1 + x^2\right)}_{\text{value-nine cards}}.$$

Expanding this polynomial using a symbolic algebra system,

$$p(x) = 1 + 4x + 9x^2 + 14x^3 + \ldots.$$

Industry Cards	Count	Location Cards	Count
Potteries	3	Belper	2
Coal Mines	3	Coalbrookdale	3
Iron Works	4	Coventry	3
Cotton Mills/Manufacturers	8	(15 Locations omitted)	
Breweries	5	Worcester	2

Figure 1.7: Industry and Location Cards in *Brass: Birmingham*.

Only 14 possible hands can be drawn from the deck at this point in the game.

The opening question about the number of possible hands in a game of *Brass: Birmingham* (p.1) can be determined using this technique. At the start of the game, each player draws eight cards from the deck of 64 cards. Figure 1.7 shows some cards in the game and their count.

There are four cards with one copy, eleven cards with two copies, seven cards with three copies, one card with four copies, one card with five copies, and one card with eight copies. This means that selecting cards from the starting deck of *Brass: Birmingham* is associated with the expression

$$p(x) = (1 + x)^4 \times (1 + x + x^2)^{11} \times (1 + x + x^2 + x^3)^7$$
$$\times (1 + x + x^2 + x^3 + x^4)^1 \times (1 + x + x^2 + x^3 + x^4 + x^5)^1$$
$$\times (1 + x + x^2 + x^3 + x^4 + x^5 + x^6 + x^7 + x^8)^1.$$

Expanding this polynomial (again, using a computer algebra system) and selecting the coefficient on x^8, which corresponds to selecting eight cards, determines that there are 6 967 695 opening hands in *Brass: Birmingham*.

Generating functions appear to calculate the correct answers as if by magic. The connection is not completely surprising as the binomial coefficients received their name from the Binomial Theorem, which describes how to expand a binomial to a power.

Binomial Theorem: For any values of x and y and any non-negative integer n,

$$(x + y)^n = \sum_k \binom{n}{k} x^k y^{n-k}.$$

Here, the summation notation $\sum_k a_k$ means to add all non-zero values of a_k together. Since $C(n, k) = 0$ for values, where k is not between 0 and n, this sum is finite as there are only $n + 1$ non-zero terms. There is even a Multinomial Theorem, presented on page 19, which extends this connection to multinomials.

Multinomial Theorem: For any values of $x_1, \ldots x_t$ and any value of $k_1, \ldots, k_t \geq 0$ with $n = k_1 + \cdots + k_t$,

$$(x_1 + \cdots + x_t)^n = \sum_{k_1 + \cdots + k_t = n} \binom{n}{k_1, \ldots k_t} x_1^{k_1} \cdots x_t^{k_t}.$$

Again, because the values of $M(k_1, \ldots, k_t) = 0$ when any $k_i < 0$, the sum is finite as there are only a finite number of non-negative values of $k_1, k_2, \ldots, k_t$, where $k_1 + \cdots + k_t = n$.

The connection between multinomial expansion and selecting objects results from the definition of polynomial multiplication. Consider the case of expanding the $(a + b + c)^5$:

$$(a + b + c)(a + b + c)(a + b + c)(a + b + c)(a + b + c).$$

Each term in the product results from selecting one term from the first multinomial, one from the second multinomial, and so forth. How many terms will be of the form $a^2 b^2 c$?

Before moving all of the as to the front and the cs to the back, an example of a product that leads to a term of the form $a^2 b^2 c$ would be *cabab*, selecting c from the first factor, a from the second factor, b from the third factor, and so forth. This is equivalent to the number of ways to shuffle a deck of five cards marked a, a, b, b, c and deal them out. This is precisely the type of counting used in the Multinomial Rule, and so there are $M(2, 2, 1)$ ways that a term of the form $a^2 b^2 c$ can be obtained.

1.5   COUNTING SUMMARY

While the basic idea of counting is relatively easy to understand, putting it into practice requires identifying the pertinent features of the situation. The four most common counting techniques that are confused with each other are quickly distinguished here. The two key features that can be used to distinguish these cases are whether the order of selection matters, together with whether the objects are only distinguishable by type or are individually distinguishable. These two conditions lead to four situations summarized in Figure 1.8.

Another feature for determining whether a counting rule applies is considering the final expression generated by the rule. In general, the factors in

	Ordered	Unordered
Individually Distinguished	Permutation Rule	Combination Rule
Distinguished by Type	Product Rule	Stars and Bars

Figure 1.8: Commonly Confused Counting Methods.

the numerator are the results of selections. In contrast, the factors in the denominator are the results of coalescing groups from the final selections into categories (or types). In its simplest form, the Product Rule has two factors and represents two selections. The Power Rule result of n^k has k factors and represents k selections. The Permutation Rule and Combination Rule result in $P(n, k)$ and $C(n, k)$, both of which have a product of k factors in their numerators. Similarly, the Multinomial Rule results in $M(k_1, \ldots, k_t)$ which has n factors in the numerator which represent the n selections. Finally, the Stars and Bars Technique formula $C(n + (t - 1), n)$ has a numerator that is the product of $t - 1$ factors representing selecting the $t - 1$ locations of the bars.

The difficulty in identifying the correct counting technique is also a reason why the formula for $C(n, k)$ was explained two different ways, and the formula for $M(k_1, \ldots, k_t)$ three different ways (with promises for more explanations to come). Why not accept the first explanation and save multiple paragraphs of the text? Mathematicians find these different connections inherently satisfying, but there are more reasons why one would want to be exposed to all of them. One of the alternate explanations may make more sense than the first one presented. If a formula needs to be reconstructed, one of these lines of reasoning may be easier to bring to mind in the process. There is also the common situation that an application of a general rule will be presented, but the presentation does not match the reasoning that led to the general rule. In these cases, knowing several descriptions of a pattern allows one to recognize it when it is described differently. This may allow one to recognize that a problem is related to a known concept. Finally, understanding how each explanation leads to the same formula helps confirm that the reasoning correctly reflects the situation. While mathematics is built on the notion of proof and deductive logic, there is still the task of recognizing which formulas apply to a particular situation. Calculating the same value with multiple methods can help increase confidence in one's intuition. This is also why Appendix A provides references to computer code to determine many of this book's results through brute force (rather than using formulas). Testing particular cases provides strong evidence that the situation was not misunderstood.

A natural question often asked is why a confusing collection of complicated formulas is created for a simple task like counting. Is there a more straightforward method that one can use that does not require careful determination of the situation or a complicated formula? The answer is that there is a more straightforward rule, Zaphod's Counting Rule.[5]

Zaphod's Counting Rule: There are always 42 ways of selecting k objects from n objects.

---

[5]Zaphod's Counting Rule is named after Zaphod Beeblebrox, a Betelgeusian celebrity who invented the Pan-Galactic Gargle Blaster, a cocktail based on Janx Spirit.

This method is much simpler; one does not have to worry about which formula to use, nor does one need to use the values of k or n. Hopefully, the issue with this more straightforward rule is clear: it frequently gives the wrong answer. Mathematicians can be described as lazy perfectionists. As perfectionists, they demand that the answer be exactly correct, or when the answer is known to be inexact, they demand error bounds that can be proven accurate. This is their overriding concern. However, given two methods that give the correct result, mathematicians will prefer the more straightforward method. The rules provided here are the simplest known methods to count objects correctly in these situations. If there were a more straightforward method that always led to the correct result, mathematicians would be using it, and it would have been included here.

1.6 LARGE NUMBERS

In the previous sections, there have been several very large numbers, such as those shown in Figure 1.9.

To put this in scale, there are roughly 10^7 seconds every year, roughly 10^{17} seconds have elapsed since the Big Bang, and a modern supercomputer (circa 2023) is capable of roughly 10^{18} operations per second. So if someone were to start a game of *Brass: Birmingham* or *Cascadia* every second for an entire year, they might see every possible opening. However, to see every opening hand of *SCOUT* and *Ark Nova* would take roughly ten million and one hundred million years, respectively. The supercomputer could list all of these starts in under one second. However, this will not approach the speed necessary to list all of the shuffles of large decks. Assuming that the supercomputer had been listing card shuffles since the Big Bang, it could only have enumerated $10^{17} \times 10^{18} = 10^{35}$ shuffles. With this perspective, it becomes clear that even listing every possible shuffle for a regular deck of cards, the tiles from *Cascadia*, or the cards from *Ark Nova* is not a reasonable task.

Count	Game Situation
Over 10^6	opening hands in *Brass: Birmingham*, p. 18
Over 10^7	opening tiles in *Cascadia*, p. 9
Over 10^8	shuffles in *Flamme Rouge*, p. 13
Over 10^{14}	opening hands in *SCOUT*, p. 10
Over 10^{15}	opening hands in *Ark Nova*, p. 12
Over 10^{67}	shuffles of a regular 52 card deck
Over 10^{128}	shuffles in *Cascadia*, p. 10
Over 10^{402}	shuffles in *Ark Nova*

Figure 1.9: Large Numbers in Tabletop Games.

Rates of Growth

Large numbers like those in Figure 1.9 will appear in several places in this book, often parameterized by some other number. For example, the number of shuffles of a deck of n cards (Section 1.2), the number of polyominoes with n squares (Section 2.3), the number of square tiles each of whose edges is one of n colors (Section 3.2) occur in the first three chapters and more are found in later chapters. In these situations, mathematicians often discuss growth rates to estimate how quickly the associated numbers grow. Saying a number is "exponentially large" does not say much about the number; for example, 1 is also an exponential number: $1 = 1000^0$. The phrase "exponentially large" often means very large, but this usage is colloquial. To say that a value grows exponentially requires that the value be compared to some benchmark. For example, the number of possibilities grows exponentially if someone makes n choices, each having two options. This is because the number of possibilities is 2^n, with the benchmark, n, appearing in the exponent of a constant. There is a great deal of literature and notation (big-O, little-o, big-Ω, little-ω, big-Θ) surrounding the study of rates of growth. Informally and sufficient for the purpose here, the notation $f(n) \prec g(n)$ means that the function $g(n)$ grows significantly faster than $f(n)$ as n increases. Relative Rates of Growth shows growth rates of common expressions in this book.

Relative Rates of Growth: For $0 < p < q$ and $1 < a < b$:

$$\underbrace{n^p \prec n^q}_{\text{polynomial}} \prec \underbrace{a^n \prec b^n}_{\text{exponential}} \prec \underbrace{n! \prec n^n \prec \ldots}_{\text{HC SVNT DRACONES}}$$

Polynomial growth is typically amenable to brute force techniques for the values of n found in tabletop games. However, even exponential growth can place problems beyond the means of brute force techniques.

It should be noted that these growth rates do not imply that the expression to the right is always greater than the expression to the left, only that the expression to the right is significantly greater if n is taken to be sufficiently large. This is shown in Table 1.1.

Returning to tabletop games, an important caveat is that merely having many starting positions does not necessarily mean the game has high replayability. There are roughly 10^{61} ways to shuffle the deck in *Candy Land*, but the game has limited replayability. On the other hand, each game of *Chess* and *Go* start with the same opening state, and both are highly replayable.

Table 1.1: Approximate Values for Common Expressions in n.

n	n^2	n^3	2^n	3^n	n!	n^n
1	1	1	2	3	1	1
2	4	8	4	9	2	4
3	9	27	8	27	6	27
$\vdots$	$\vdots$	$\vdots$	$\vdots$	$\vdots$	$\vdots$	$\vdots$
10	100	1000	1024	59049	10^6	10^{10}
$\vdots$	$\vdots$	$\vdots$	$\vdots$	$\vdots$	$\vdots$	$\vdots$
100	1×10^4	1×10^6	1×10^{30}	5×10^{47}	9×10^{157}	10^{200}

CHAPTER 2

Geometry

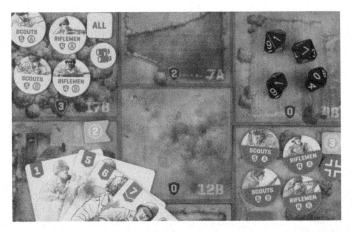

Figure 2.1: Components from *Undaunted: Normandy*.[1]

Undaunted: Normandy is a two-player game set shortly after the World War II landings in Normandy. Players control small squads of soldiers by drawing and then playing cards for each squad member. Movement by members of the squads is done on an offset square tiling, as shown in Figure 2.1. The game *Undaunted: Stalingrad* uses the same system, while *Undaunted: Battle of Britain* uses a hexagonal tiling.

Why are square and hexagonal tilings common in tabletop games, while pentagons, heptagons, and octagons are uncommon?

---

[1] Undaunted: Normandy by David Thompson & Trevor Benjamin ©2021 Osprey Publishing Ltd.

 DOI: 10.1201/9781003383529-2

INTRODUCTION

While Chapter 1 focused on counting, this chapter will focus on geometry. When a tabletop game involves a tiling (a grid), these tilings are always composed of triangles, squares, and hexagons. Euclidean geometry can be used to demonstrate that these three tilings are the only tilings available to game developers, as the tabletop's limitations preclude more exotic tiles (such as pentagons, heptagons, or octagons).

Variations on these tilings are introduced, the first of which is to offset the tiles as shown in *Undaunted: Normandy*. The second variation uses polyominoes (tiles with shapes like those in Tetris) to increase the possibilities. Polyominoes lead to questions about the symmetry of shapes that can be rotated and reflected (flipped), a topic taken up in Chapter 3.

A common theme in mathematics after discovering a restriction is generalizing to situations, where the restriction is lifted. While this exploration has little impact on tabletop game design (as their boards need to fit on a tabletop), it does lead to some interesting mathematics. In particular, pursuing this idea identifies the five Platonic solids and introduces hyperbolic geometry.

2.1 REGULAR TILINGS

Several games use tilings based on triangles, squares, and hexagons (Figure 2.2). Even in games where individual tiles are not these shapes, the tiles are often compatible with these three tilings. This section will discuss why other tilings are rarely used for tabletop games.

An *n*-gon is a polygon with *n* sides. When *n* is small, these shapes have specialized names (such as triangle, quadrilateral, pentagon, and so forth). The sides are referred to as edges, and the places where two edges of a polygon meet are referred to as vertices, singular vertex. At each vertex, the two edges will form an angle on the inside of the polygon, which is an interior angle. A regular polygon is one whose edges are the same length, and interior angles are the same size.

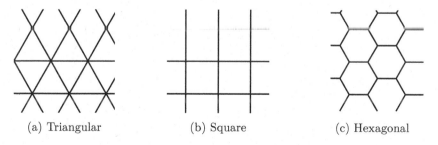

(a) Triangular   (b) Square   (c) Hexagonal

Figure 2.2: Regular Edge-to-Edge Tilings.

(a) Triangular   (b) Square

Figure 2.3: Regular Offset Tilings.

A regular tiling is a collection of regular polygons that fill the plane without any gaps. Initially, edges and vertices are required to coincide in the tiling as in Figure 2.2 (such tilings are sometimes referred to as edge-to-edge tilings). This excludes the possibility of the tilings shown in Figure 2.3, known as offset tilings. Offset tilings are discussed in Section 2.2 because offset tilings are also used in games. If two tiles share any part of an edge other than a vertex, they are adjacent, sometimes the term edge-adjacent will be used for clarity.

Every regular n-gon can be divided into n triangles by extending radii from the center of the n-gon to the vertices (see Figure 2.4a for the example of a hexagon). Because the polygon is regular, all the central angles are equal, and the variable α will represent their size. Similarly, all the polygon's interior angles are equal, and the variable β will represent their size. Note that the triangle created by the perimeter and the radii will have angles of size $\beta/2$ at the perimeter (see Figure 2.4a).

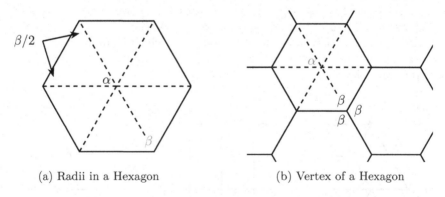

(a) Radii in a Hexagon   (b) Vertex of a Hexagon

Figure 2.4: Angles in a Regular Polygon Tiling.

For a regular planar n-gon, the size of the central angle α will equal $360°/n$, and the sum of the angles in any triangle will be $180°$, so

$$\frac{\beta}{2} + \frac{\beta}{2} + \alpha = 180°,$$

$$\beta + \frac{360°}{n} = 180°. \tag{†}$$

Now consider a corner in the tiling where k polygons meet at a vertex (see Figure 2.4b for the case of three hexagons meeting at a vertex). Because each interior angle is β, $k\beta = 360°$ or $\beta = 360°/k$. When substituted into Equation (†), the resulting equation gives a necessary condition for a regular plane tiling.

$$\frac{360°}{k} + \frac{360°}{n} = 180°$$

Dividing this equation by $360°$ arrives at the Regular Tiling Condition.

Regular Tiling Condition: If a planar edge-to-edge plane tiling consists of regular n-gons meeting k at each vertex, then

$$\frac{1}{k} + \frac{1}{n} = \frac{1}{2}$$

It is possible to consider all integers n and k for which $1/n + 1/k = 1/2$ to determine all possible edge-to-edge tilings of the plane with regular polygons. Table 2.1 presents the values of this sum for n and k less than eight.

Notice that if either n or k is two, then $1/n + 1/k > 1/2$. If n and k are both greater than two and at least one of them is greater than six, then $1/n + 1/k < 1/2$. So the table contains all of the values where $1/n + 1/k = 1/2$. There are only three possible tilings: hexagons which have $n - 6$ sides with

Table 2.1: Regular Tiling Candidates: Values of $1/n + 1/k$ for $2 \leq n, k \leq 7$.

n		k					
	2	3	4	5	6	7	$\cdots$
2	1	$5/6$	$3/4$	$7/10$	$2/3$	$9/14$	$\cdots$
3	$5/6$	$2/3$	$7/12$	$8/15$	$1/2$	$10/21$	$\cdots$
4	$3/4$	$7/12$	$1/2$	$9/20$	$5/12$	$11/28$	$\cdots$
5	$7/10$	$8/15$	$9/20$	$2/5$	$11/30$	$12/35$	$\cdots$
6	$2/3$	$1/2$	$5/12$	$11/30$	$1/3$	$13/42$	$\cdots$
7	$9/14$	$10/21$	$11/28$	$12/35$	$13/42$	$2/7$	$\cdots$
$\vdots$	$\vdots$	$\vdots$	$\vdots$	$\vdots$	$\vdots$	$\vdots$	$\ddots$

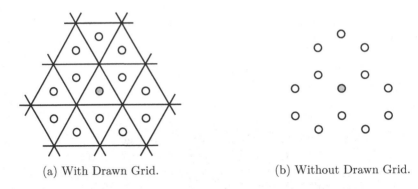

(a) With Drawn Grid.   (b) Without Drawn Grid.

Figure 2.5: Vertex-Adjacency in Triangular Tiling.

$k = 3$ hexagons meeting at each vertex; squares, which have $n = 4$ sides with $k = 4$ squares meeting at each vertex; and triangles, which have $n = 3$ sides with $k = 6$ triangles meeting at each vertex. These are precisely the three tilings shown in Figure 2.2.

Triangular Tilings

Triangular tilings are not widely used in tabletop games, as has been noted in online forums. However, these tilings are not completely absent. For example, *Tri-Ominos*, a variation on *Dominoes*, is an example of a game that uses a triangular tiling.

The most obvious disadvantage of a triangular tiling is it limits the number of adjacent tiles, as each tile in a triangular tiling is only adjacent to three tiles. This is more restrictive than square tilings (four adjacent tiles) and hexagon tilings (six adjacent tiles). This could be increased by considering two tiles to be adjacent if they share a vertex (referred to as being vertex-adjacent). However, there is a significant variation in the distance between vertex-adjacent tiles. In Figure 2.5a, the centers of thirteen tiles are marked, and all tiles marked with an open circle are vertex-adjacent to the central tile (marked with a gray-filled circle). Removing the gridlines (in Figure 2.5b) shows a wide discrepancy between the distances between adjacent tiles.

To determine the amount of discrepancy, begin with a regular triangle with a side length of two and draw bisectors from each angle. These three bisectors all meet in the center of the triangle and make a right angle with the opposite side at its midpoint (see Figure 2.6, which includes separate drawings of the two relevant subfigures). Using an overline to indicate the size of a line segment, in $\triangle ABC$, $\overline{BC} = 2$, $\overline{AC} = 1$ and the Pythagorean Theorem[2] determines $\overline{AB} = \sqrt{4 - 1} = \sqrt{3}$.

---

[2]Pythagoras of Samos was a sixth-fifth century BCE Greek philosopher who may or may not have discovered the Pythagorean Theorem.

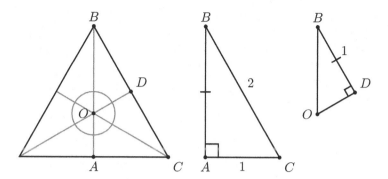

Figure 2.6: Geometry of Regular Triangle.

Because $\triangle ABC$ and $\triangle DBO$ are both right triangles that share the angle at B, the triangles are similar with side AB corresponding to side DB (shown with the mark). The ratio of the corresponding sides of two similar triangles is constant, and the ratio between sides AB (with length $\sqrt{3}$) and DB (with length 1) is $1/\sqrt{3}$. Scaling all the relevant sides of $\triangle ABC$ by this value determines the sides of $\triangle DBO$:

$$\overline{OD} = \frac{1}{\sqrt{3}}\overline{CA} = \frac{1}{\sqrt{3}},$$
$$\overline{OB} = \frac{1}{\sqrt{3}}\overline{CB} = \frac{2}{\sqrt{3}}.$$

With the pertinent lengths computed, assume the triangles in Figure 2.5a have a side of length two. The distance between the centers of two triangles that share an edge will be $2\overline{OD} = 2/\sqrt{3}$. In comparison, the distance between the centers of two triangles directly across a vertex from each other will be $2\overline{OB} = 4/\sqrt{3}$. This means that the physical distance between adjacent centers can vary by up to a factor of two:

$$\frac{\text{maximum distance between vertex-adjacent tiles}}{\text{distance between edge-adjacent tiles}} = \frac{4/\sqrt{3}}{2/\sqrt{3}} = 2.$$

Requiring movement through the edges of each triangle keeps the distance between the centers of all edge-adjacent tiles constant. However, it does not entirely solve the discrepancy between the distance measured by a ruler and the distance measured by steps. For example, in Figure 2.7a, to move from the triangle with the gray circle to the triangle directly above it requires 3 steps for a total length of $6\sqrt{3}$ even though the direct distance between the centers of these triangles is only $4\sqrt{3}$. The walk from the lower triangle to the triangle above gives the impression of someone lurching right and left as they attempt to walk in a straight line and has been referred to as a drunken walk. This drunken walk behavior imposes a penalty for movement in some directions,

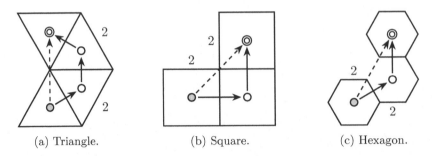

(a) Triangle.   (b) Square.   (c) Hexagon.

Figure 2.7: Drunken Walks in Regular Tilings.

and for a triangle, there can be up to a 50% penalty:

$$\frac{\text{distance via adjacent triangles}}{\text{direct distance}} = \frac{6\sqrt{3}}{4\sqrt{3}} = 1.50.$$

Another issue with triangular tilings is the size of the interior. The small angles at each triangle vertex result in less usable space in the center of the tile, away from the perimeter. This can be important when multiple pieces should be placed in a single tile, as pieces in these angular regions are easily jostled into another tile. One way to estimate the size of this effect is to consider the percentage of the region closer to the center than the perimeter. Determining the exact shape of the region requires more advanced mathematics, but it can be estimated by considering the circle whose radius is one-half the radius of the inscribed circle. This circle is drawn in Figure 2.6. Points inside the circle are all central, while those outside are more peripheral. The circle is shown in Figure 2.6 for a regular triangle. The circle will have a radius of $\sqrt{3}/6$ and area of $\pi \left(\sqrt{3}/6\right)^2 = \pi/12$ while the area of the triangle will be $(1/2)(2)(\sqrt{3}) = \sqrt{3}$. This means that the inscribed circle will enclose approximately 15% of the area of the entire triangle:

$$\frac{\text{area of circle}}{\text{area of triangle}} = \frac{\pi/12}{\sqrt{3}} = \frac{\pi}{12\sqrt{3}} \approx 0.15.$$

The exact value for the percentage of points closer to the center than the perimeter is $5/27 \approx 19\%$ (see Appendix A).

Square Tilings

Square tilings occur frequently in tabletop games as they match the typical notion of choosing a direction: up/down/right/left or north/south/east/west. They offer a reasonable amount of adjacency: four adjacent tiles and eight vertex-adjacent tiles. In the tabletop gaming community, edge-adjacent squares are often referred to as "orthogonally adjacent," while vertex-adjacent

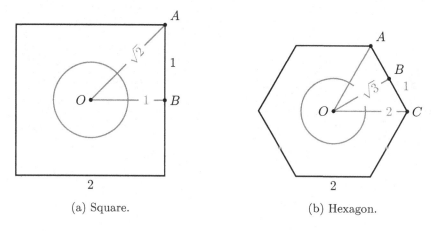

(a) Square.　　　　　　　　　　(b) Hexagon.

Figure 2.8: Geometry of Squares and Regular Hexagons.

squares are said to be "diagonally adjacent." The geometry of a square tile is much simpler than that of a triangular tile, and the pertinent distances can be found in Figure 2.8a. The Pythagorean Theorem is enough to obtain the one missing value $\overline{OA} = \sqrt{1+1} = \sqrt{2}$.

Like triangular tilings, the distances between the centers of vertex-adjacent tiles can differ; however, the longer distance between tile centers is only about 41% longer than the shorter distance between tile centers:

$$\frac{\text{maximum distance between vertex-adjacent tiles}}{\text{distance between edge-adjacent tiles}} = \frac{2\sqrt{2}}{2} \approx 1.41.$$

So, the effect is less extreme than in the triangular case. The drunken walk problem is also not quite as extreme (see Figure 2.7b), with only a 41% penalty:

$$\frac{\text{distance via adjacent squares}}{\text{direct distance}} = \frac{4}{2\sqrt{2}} \approx 1.41.$$

Finally, in a square tiling, the central circle in a square will take up roughly 20% of the available space:

$$\frac{\text{area of circle}}{\text{area of square}} = \frac{\pi(\frac{1}{2})^2}{(2)^2} \approx 0.20.$$

The exact value is $(4\sqrt{2} - 5)/3 \approx 22\%$.

There are practical reasons for choosing square tiles. For example, they fit conveniently onto punchboards (which tend to be rectangular) with little wasted space, and they may also be easier to match with artwork, especially artwork containing text (see David Thompson's comment on page 36). Chess uses a square tiling and is shown in Figure 2.9a.

(a) Square Tiling in *Chess.*

(b) Hexagonal Tiling in *Hoplomachus: Remastered.*

Figure 2.9: Regular Tilings in *Chess* and *Hoplomachus: Remastered*[3].

Hexagonal Tilings

Hexagonal tilings are less common in everyday activities. This can be seen in the fact that hexagonal directions are derived from the directions of a square tiling (northwest, north, northeast, southeast, south, and southwest). In addition to allowing the maximum number of adjacent tiles through edges, they have the property that any two tiles that share a vertex will also share an edge (so every vertex-adjacent tile is also edge-adjacent). The geometry of a regular hexagon can be determined by splitting the hexagon radially into six regular triangles (see Figure 2.8b). Because $\triangle ACO$ is a regular triangle of with side length of two, $\overline{OC} = 2$ and $\overline{OB} = \sqrt{3}$.

Because no two hexagonal tiles are adjacent only through a vertex, there is no variation in the distance between the centers of vertex-adjacent hexagons. Like triangles and squares, hexagons exhibit the drunken walk phenomenon, but it is much less extreme (see Figure 2.7c). The total displacement will be $4\sqrt{3}$. However, the distance between the two centers is only six. This gives a movement penalty of approximately 15%:

$$\frac{\text{distance via adjacent squares}}{\text{direct distance}} = \frac{4\sqrt{3}}{6} \approx 1.15.$$

Finally, the central circle will take up approximately 23% of the hexagon's area:

$$\frac{\text{area of circle}}{\text{area of hexagon}} = \frac{\pi(\frac{\sqrt{3}}{2})^2}{6(^1/_2)(2)(\sqrt{3})} \approx 0.23.$$

The exact value is $(16 - 9\sqrt{3})/\sqrt{3} \approx 24\%$.

---

[3]Hoplomachus: Remastered and Hoplomachus Victory images and text were used with permission from Chip Theory Games. They are copyrighted by Chip Theory Games, and all rights are reserved worldwide.

Many games from a war game background, such as *Undaunted: Battle of Britain* (p.24), use hexagonal tilings, as well as games that mimic small-scale skirmishing, such as *Gloomhaven* (p.84), *Oathsworn: Into the Deep Wood* (p.86), and *Hoplomachus: Remastered* (p.85) (shown in Figure 2.9b). This is because hexagonal tilings provide more natural movement for the combatants.

Dual Tilings

While the dots in Figure 2.5a show the centers of thirteen triangles, the same dots in Figure 2.5b also clearly show the vertices of a hexagonal grid. In general, every tiling has a dual tiling defined as the tiling whose vertices are the centers of the original tiling with edges connecting the centers of adjacent tiles in the original tiling.

Starting with a hexagonal tiling in black in Figure 2.10a, its dual tiling is dashed and gray. Notice that the dual of a hexagonal tiling is a triangular tiling. Repeating the process (finding the dual of the dual of a tiling) will return to the original tiling (the term dual is commonly used in mathematics for things with this property). So, the dual of a triangular tiling is a hexagonal tiling, and the dual of a hexagonal tiling is a triangular tiling. This means that whenever a designer wants to use triangular tilings, they could use hexagonal tilings with placement at the vertices. For regular tilings of n-gons with k tiles meeting at a vertex, the dual regular tiling will be a tiling of k-gons with n tiles meeting at a vertex. Because $n = 4$ and $k = 4$ in a square tiling, a square tiling is self-dual, meaning that the dual of a square tiling is again a square tiling (see Figure 2.10b).

Watergate is a two-player game where one player takes on the role of Nixon, and the other takes on the role of the Washington Post reporters uncovering the story. In the game, both players collect evidence. The Nixon player will obscure the evidence to block the path between him and other co-conspirators, while the reporters will use the evidence to link Nixon to the co-conspirators. Both players achieve their objectives by placing the evidence on the vertices of a triangular tiling, as shown in Figure 2.11a.

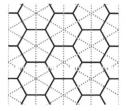

(a) Triangular and Hexagonal.

(b) Square.

Figure 2.10: Dual Tilings.

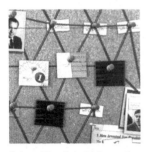
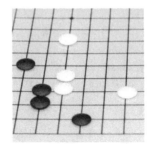

(a) Dual Triangular Tiling in *Watergate*.

(b) Dual Square Tiling in *Go*.

(c) Dual Hexagonal Tiling in *Catan*.

Figure 2.11: Dual Tilings in *Watergate*[4], *Go*, and *Catan*[5].

The classical game of *Go* is played by placing pieces on a square tiling, as shown in Figure 2.11b. Using the dual tiling here clarifies that only pieces placed orthogonally are relevant.

Catan has players take the role of settlers who build settlements and roads on an island. Settlements are placed on the vertices of a hexagonal tiling, as shown in Figure 2.11c, where each settlement can obtain resources from the three adjacent tiles.

2.2 OFFSET REGULAR TILINGS

In the previous section, the vertices of adjacent tiles coincided. In this section, the vertex of one tile may be on the edge of another tile as in Figure 2.3. To what extent does this increase the number of available tilings?

In this situation, an analysis similar to Section 2.1 can be done. Let k be the number of tiles sharing a vertex. For example, Figure 2.12a shows a single vertex where $k = 2$ since the vertex is a vertex of the two tiles above the horizontal line. Notice that three tiles are sharing this point since the tile below the line shares the point but is not counted since it is not a vertex of the lower tile. Similarly, Figure 2.12b shows the case where $k = 3$. Because the vertex occurs on the edge of a tile, the k angles must sum to $180°$:

$$k\beta = 180°,$$

$$\beta = \frac{180°}{k}.$$

[4]Watergate images and text were used with permission from Capstone Games. They are ©2019 Capstone Games, ©2019 Frosted Games. All rights are reserved worldwide.

[5]Catan images and text were used with permission and are copyright ©1995 CATAN GmbH - CATAN, the CATAN logo, the "CATAN Sun," and the CATAN Brand Logo are trademark properties of CATAN GmbH (catan.com). All rights reserved.

(a) Two Tiles at a Vertex.　　　　　(b) Three Tiles at a Vertex.

Figure 2.12: Angles in Offset Polygon Tiling.

Combining this with Equation (†) on page 27 yields the Offset Tiling Condition.

Offset Tiling Condition: If a planar offset tiling consists of regular n-gons meeting k at each vertex, then

$$\frac{1}{2k} + \frac{1}{n} = \frac{1}{2}.$$

Again, an exhaustive search can be performed for small values of n and k. The results are shown in Table 2.2. The two possible values for n and k are precisely those shown in Figure 2.3, where three triangles meet at a vertex and where two squares meet at a vertex.

Furthermore, if the game only focuses on adjacency, there is no more variety than the regular tilings. For example, the offset triangular tiling has the same adjacencies as a regular square tiling, and the offset square tiling has the same adjacencies as a hexagonal tiling. Figure 2.13 shows the process of transforming the offset triangular tiling into a regular square tiling. A similar process shows the offset square tiling is equivalent to the hexagonal tiling.

Using offset tilings has some downsides. In many games, players must place tiles adjacent to previous tiles, often with some requirement of matching edges. Having one edge of a tile adjacent to an entire edge of an adjacent tile in some cases and adjacent to edges of two adjacent tiles in other cases significantly increases the complexity.

Table 2.2: Offset Tiling Candidates: Values of $1/2k + 1/n$ for $2 \leq n \leq 5, 2 \leq k \leq 4$.

n	k			
	2	3	4	$\cdots$
2	$3/4$	$2/3$	$5/8$	$\cdots$
3	$7/12$	$1/2$	$9/20$	$\cdots$
4	$1/2$	$5/12$	$3/8$	$\cdots$
5	$9/20$	$11/30$	$13/40$	$\cdots$
$\vdots$	$\vdots$	$\vdots$	$\vdots$	$\ddots$

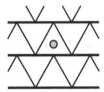 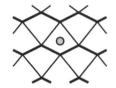

Figure 2.13: Transforming an Offset Tiling to a Regular Tiling.

Offset tilings may be appropriate when these mechanisms are not being used. For example, *Undaunted: Normandy* (p.24) uses an offset square tiling with a typical game setup shown in the opening figure of this chapter. When asked about this design choice, David Thompson, designer of *Undaunted: Normandy*, said the following:

> "The use of offset squares was one of practicality and aesthetics. I wanted Undaunted to appeal to a demographic that straddled both wargamers and non-wargamers, and I felt offset squares *might* have broader aesthetic appeal. Certainly I think it can be easier to have art that fits better within a square than a hex.
>
> I also benefited from the fact that the elements that are better suited for hexes (facing, directional attacks, and line of sight) aren't included in Undaunted.
>
> So essentially, because they are nearly identical from a topological standpoint, and the strengths of hexes weren't relevant, I felt comfortable going with offset squares."
>
> —David Thompson [51]

2.3 POLYOMINO TILINGS

One way to add variety to tile shapes is to continue using regular tilings but to use shapes consisting of multiple tiles in that tiling. The most common of these are polyominoes. A polyomino is a figure created by any number of same-sized squares connected edge-to-edge. They are often categorized by the number of squares in the shape. For example, a monomino (one-square), two biominoes (two squares), and two triominoes (three squares) appear in Figure 2.14.

Many polyomino games involve players placing polyomino-shaped tiles onto a regular grid. Unlike regular tilings (where all tiles are the same shape), whether two polyominoes are equivalent is often a critical game feature. For example, which size four polyominoes (called tetrominoes) in Figure 2.15 should be considered the same?

None of the polyominoes in the top row (polyominoes (a), (b), and (c)) are equivalent to any other polyomino in the figure. However, the situation with the polyominoes on the bottom row is more complicated and depends on the game. Polyomino (d) can be rotated to match polyomino (e), which

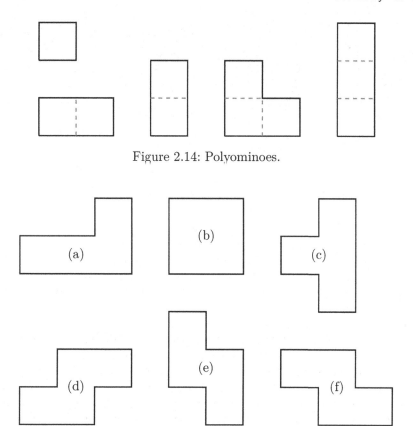

Figure 2.14: Polyominoes.

Figure 2.15: Tetrominoes.

would be allowed in most polyomino games. When rotation is *not* allowed, the polyomino is a fixed polyomino. Fixed here refers to the fixed orientation of the polyomino, not to being fixed at a particular location. On the other hand, to match polyomino (d) to polyomino (f) requires the polyomino to be reflected (flipped). When rotation is allowed, and reflection is *not* allowed, the polyomino is a one-sided polyomino. When a polyomino can be rotated and reflected, it is a free polyomino.

In *My City*, players place polyomino-shaped tiles representing city buildings on their player boards, which represent their cities. Throughout the game's 24 scenarios, the placement and scoring conditions are varied so that playing through the scenarios provides a good representation of the types of ways polyomino games are played. In *My City*, tiles can be rotated but not flipped, so they are one-sided polyominoes. Figure 2.16a and 2.16b show the front and back of the *My City* tiles. The fact that flipping tiles is not allowed in gameplay is reflected by the fact that the reverse sides do not display reflected versions of the front.

(a) Front of tile from *My City*. (b) Back of tile from *My City*. (c) Front of tile from *The Isle of Cats*. (d) Back of tile from *The Isle of Cats*.

Figure 2.16: Polyomino Tiles in *My City*[6] and *The Isle of Cats*[7].

In *The Isle of Cats*, players score victory points by placing polyomino tiles representing cats onto a boat to save the cats from the evil Lord Vesh. Figure 2.16c and 2.16d show the two sides of *The Isle of Cats* tiles. The tile indicates that flipping is allowed by having the reverse side display a reflected image, so these are free polyominoes.

Tiling regions with polyominoes and counting polyominoes of a specified size have received the attention of mathematicians and computer scientists. Counting polyominoes is computationally challenging. Even as of 2023, researchers have determined the number of polyominoes for sizes only up to around 50 squares. However, the task of counting for larger sizes is computationally challenging because the number of polynomials grows exponentially, with the number of polyominoes of size n approximately equal to λ^n, where $4.00253 \leq \lambda \leq 4.5252$.

Although less common in tabletop games, the same construction can be used with the triangular and hexagonal regular tilings. The triangular version of a polyomino is referred to as a polyiamond. The "iamond" comes from the word diamond, which is the shape obtained by adjoining two regular triangles along one edge. An example of a game that uses tiles composed of adjoined regular triangles is *Hamlet: The Village Building Game*, see Figure 2.17. In this game, players lay out tiles representing buildings and resources in a common area. As the game progresses, players will score points by contributing to the development of the hamlet and the construction of a church (which elevates the hamlet to a village).

The hexagonal version of a polyomino is a polyhex. In *Ark Nova* (p.11), players work to develop their zoo by placing polyhex tiles onto their player board to represent enclosures for the animals in their zoo. The tile is flipped when an animal is placed in an enclosure, so the enclosure tiles must be

---

[6]"My City" by Reiner Knizia, Michael Menzel ©2020 Franckh-Kosmos Verlags GmbH & Co. KG, Stuttgart, Germany. "My City" by Reiner Knizia, Michael Menzel ©2020 Thames & Kosmos, LLC, Providence, RI, USA

[7]Images and photographs of The Isle of Cats are copyrighted to The City of Games Limited.

Figure 2.17: Polyiamonds in *Hamlet: The Village Building Game.*[8]

(a) Tile Layout in *Ark Nova.*　(b) Tile Layout in *Ark Nova.*　(c) Tile Layout in *Ark Nova.*　(d) Tile Layout in *Ark Nova.*

Figure 2.18: Polyhexes in *Ark Nova.*[9]

the same shape when reflected (see Figure 2.18a and 2.18b). Some special enclosures are one-sided and must be placed in the zoo with a particular side facing up; these tiles are generally asymmetrical and have a distinct front and back (as shown in Figures 2.18c and 2.18d).

2.4 REGULAR TILINGS OF SURFACES

In the Regular Tiling Condition, it was shown that a plane can be tiled by n-gons meeting k at a vertex exactly when $1/n + 1/k = 1/2$. This section will explore values of n and k that do not satisfy this condition.

Consider the portion of Table 2.1 for which $1/n + 1/k > 1/2$ and $n > 2$, $k > 2$. Figure 2.19a shows the case when $n = 3$, $k = 3$: three regular triangles around vertex A. At this point, the three 60° angles at vertex A are insufficiently large to surround the vertex in the plane. If the figure can extend into three dimensions, the vertex can be surrounded by three triangles, as shown in Figure 2.19b. This type of construction will be referred to as gluing the side AB to the side AB'. This creates the cone (three of the four sides of the

---

[8]Hamlet images used with permission from owners: Mighty Boards/David Chircop.
[9]Ark Nova images used with permission from Feuerland Spiele. Ark Nova is ©2021 Feuerland Spiele, and all rights are reserved worldwide.

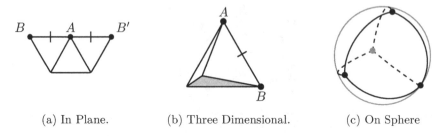

(a) In Plane.   (b) Three Dimensional.   (c) On Sphere

Figure 2.19: Tetrahedron as Spherical Tiling.

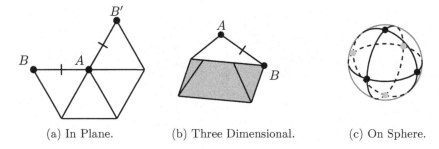

(a) In Plane.   (b) Three Dimensional.   (c) On Sphere.

Figure 2.20: Octahedron as Spherical Tiling.

tetrahedron) shown in Figure 2.19b. Placing a final tile on the bottom of the figure creates a regular polyhedron consisting of triangles meeting three at each vertex.

The shortest distance between two points on a sphere consists of an arc of a great circle (a great circle is a circle whose center coincides with the sphere's center). A polygon on the sphere consists of the region between arcs of great circles, and, on the sphere, the sum of angles in a triangle is always greater than 180°. As a result, a vertex can be adjacent to fewer triangles than is possible in the plane. Evenly spacing the four vertices of the tetrahedron on a sphere and connecting them with arcs of great circles produces Figure 2.19c. Tilings on a sphere whose edges are arcs of great circles are referred to as spherical tilings. Figure 2.19c shows a tiling of the sphere with three triangles meeting at every vertex. Even though some of the edges between vertices or some of the angles at the vertices may appear to be of different sizes, they are not. This distortion results from projecting the sphere onto the page (a plane). Each angle is the same size, 120°, and each edge is the same length (which depends on the sphere's radius).

Repeating the process with $n = 3$ and $k = 4$ also produces a spherical tiling, as shown in Figure 2.20. From the cone (the four triangular sides of a pyramid) in Figure 2.20b, two more triangles must be attached to each vertex

Figure 2.21: Spherical Tilings using Digons.

Figure 2.22: Spherical Tilings using Two n-gons.

at the base of the cone. This can be accomplished by gluing a second cone to the bottom of the original cone. This completes an octahedron, which can be placed on a sphere as shown in Figure 2.20c.

Similar constructions can be done for $n = 3$, $k = 5$ resulting in a icosa-hedron, $n = 4$, $k = 3$ resulting in a cube, and $n = 5$, $k = 3$ resulting in a dodecahedron. These values represent regular tilings of the sphere and are the Platonic solids and the most common polyhedral dice used in tabletop games. Considering spherical regions instead of planar regions gives an interpretation of a digon (a polygon with two edges): a region on the sphere with the shape of the rind of an orange wedge. Figure 2.21 shows tilings of the sphere with three and four digons.

Another feature of a sphere is that the tiles curve outward from the sphere's center. This allows two different tiles to share their boundaries. As a result, an infinite number of tilings with $k = 2$ are also possible on the sphere. This is done by evenly spacing n vertices around the equator with the northern and southern hemispheres each consisting of an n-gon. Figure 2.22 shows two possibilities: 3-gons meeting two at a vertex and 4-gons meeting two at a vertex.

Hyperbolic Tilings

To find heptagons and octagons, the other portion of Table 2.1, where $1/n + 1/k < 1/2$, must be examined. This portion is much larger and more complicated. Consider $n = 3$ and $k = 8$ (choosing $k = 8$ will make some arithmetic slightly simpler than $k = 7$). In the spherical case, the outer edges of the shape in Figure 2.19a were not long enough to allow vertices B and B' to coincide. In the hyperbolic case, the reverse occurs as the perimeter of the

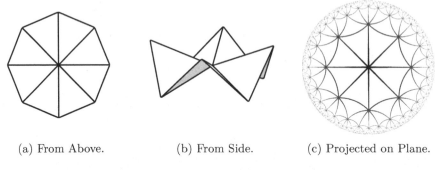

(a) From Above.   (b) From Side.   (c) Projected on Plane.

Figure 2.23: Hyperbolic Tiling [10].

eight regular triangles around a single vertex will result in a longer perimeter than will fit in the plane.

To understand how to represent this, return to the polygonal tetrahedron in Figure 2.19b. In this case, the lines and angles are all the same size but are distorted by perspective. Doing the same for the case of eight triangles in the hyperbolic space will provide a projection of hyperbolic space. When viewed from above (in Figure 2.23a), the triangles appear non-regular (having a smaller angle at the vertex). However, this is just a matter of perspective. Changing perspective to view the object from the side (in Figure 2.23b) shows that eight regular triangles have been rotated in three-space to surround a single vertex.

In the spherical case, this expansion eventually closed the shape into a polyhedron with a finite number of triangles. However, in the hyperbolic case, the need to rotate each triangle around the vertex will cause the shape to curve back on itself, but the edges will not match up as they did in the spherical case. As a result, the tilings of hyperbolic surfaces are infinite. It can be proven that the shape can be constructed without self-intersection in four-space, but it cannot be constructed in three-space. No physical games involve hyperbolic tilings because hyperbolic surfaces cannot be created in three-space. Video games are not constrained in this way, and *Hyperbolica* [34] is a virtual game played in hyperbolic space.

However, to finish the topic, it is possible to project the hyperbolic surface onto the plane to view a two-dimensional representation of hyperbolic space. When spheres project onto disks in the plane, each point in the disk's interior is the image of two points on the sphere, and great circles project onto ellipses. When hyperbolic space projects onto disks in the plane, each point in the disk's interior is the image of exactly one point on the hyperbolic space. However, hyperbolic lines project onto diameters or half-circles that meet the edge of the disk orthogonally. The tiling with eight triangles meeting at each vertex

---

[10]Figure 2.23c is public domain, credited to Parcly Taxel available at `https://en.wikipedia.org/wiki/Order-8_triangular_tiling\#/media/File:H2-8-3-primal.svg`

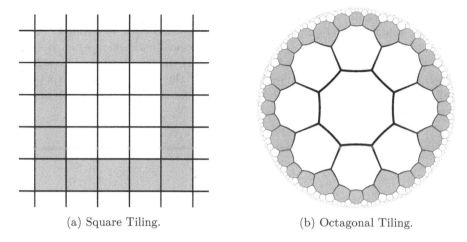

(a) Square Tiling.　　　　　　　(b) Octagonal Tiling.

Figure 2.24: Perimeter Lengths in the Euclidean and Hyperbolic Planes[11].

is shown in Figure 2.23c. In hyperbolic space, each edge is straight and of the same length, and each angle is the same size. Furthermore, every vertex is equally central in hyperbolic space in the same way that every point is equally central in a sphere. Points that appear "closer" to the boundary only appear that way due to the distortion caused by the projection.

One can convert the tiling of triangles meeting eight at a vertex to a tiling of octagons in the same way six triangles can be combined in the regular triangular tiling to form a hexagonal tiling. Some suggest that octagonal tilings can be simulated using a square tiling with vertex-adjacency. While this does result in eight adjacent tiles to the center tile, it is not the same as an actual octagonal tiling. The best way to see this is to consider the number of tiles exactly two steps from the center tile. In the case of a square tiling with vertex-adjacency, there are 16 tiles two steps from the center tile, as shown in Figure 2.24a. However, in an octagonal tiling (where three octagons meet at a vertex), there are 32 tiles that are two steps away from the center tile, as shown in Figure 2.24b. Considering squares on the sphere, the spherical square tiling (a cube) has only one tile two steps away from the center tile, namely the opposite face.

This is consistent with the previous comments about the perimeters of circles in these spaces. For spherical space, the perimeter will remain bounded. For a circle centered at the north pole, as the circle's radius grows, the circle's perimeter will initially grow until it reaches the equator. As the radius continues to grow, the circle's perimeter will now shrink until the circle reaches the south pole (where the perimeter is reduced to zero). At this point, it will

---

[11]Figure 2.24b is public domain, credited to Parcly Taxel available at https://en.wikipedia.org/wiki/Octagonal_tiling\#/media/File:H2-8-3-dual.svg

start to grow again. The formula for the perimeter of a circle of radius r on the unit sphere is $C = 2\pi \sin(r)$. Interesting, when $\pi < r < 2\pi$, $C < 0$. This indicates that the circle is traced in the opposite direction when viewed from the north pole. For a plane, the formula is the standard one, $C = 2\pi r$, which grows linearly. For a hyperbolic space,[12] the circle will grow exponentially, $C = 2\pi \sinh(r)$, where sinh is the hyperbolic sine function, which grows exponentially.

---

[12]Technically, a hyperbolic space with constant curvature of -1.

CHAPTER 3

Group Theory

Figure 3.1: Components from *Galaxy Trucker*.[1]

In *Galaxy Trucker*, players draw and play tiles onto their player boards to build spaceships. Each tile can have one of four edge types, and to place a tile, the edges of the placed tile must be compatible with the adjacent tiles. The interiors of the tiles represent different ship components, such as cargo bays, crew quarters, laser cannons, energy stores, protective shields, and engines. Some of these components have restrictions on their orientation, while others do not. The players are allowed to rotate their tiles before placing them. Two tiles are distinct if one cannot be rotated to match the other, so the energy stores shown Figure 3.1 are distinct.

How many distinct tiles are possible in *Galaxy Trucker*?

---

[1]Galaxy Trucker images used by permission from Czech Games Edition. Galaxy Trucker is ©2007 Czech Games Edition, and all rights are reserved worldwide.

DOI: 10.1201/9781003383529-3

INTRODUCTION

This chapter combines the two themes from the previous chapters: counting and tiling. The games of interest here are often described as tile-placement games. Players' moves consist of placing tiles into a grid. In some games, like *Galaxy Trucker* (p.45), players place tiles into a personal grid, while in other games, like *Carcassonne* (p.46), all players work on the same grid of tiles.

In tile-placement games, the imagery on the tile is significant to gameplay, either because tile edges must match the adjacent tiles' edges or because such matching is encouraged by providing more points. To facilitate play, players must be allowed to rotate (and possibly reflect) the tile they are placing into the grid. This ability to manipulate the tile before placing it complicates the counting of tiles, as two tiles that appear different may just be rotated or reflected copies of each other.

To handle these situations, the theory of groups is introduced. This theory provides a way to formalize the rotations and reflections available to the player and allows for accurately counting objects that may be transformed. The chapter ends by returning to a topic from Chapter 1 and providing another derivation of the binomial and multinomial formulas.

3.1 GROUPS

A more straightforward tiling question arises in the game *Carcassonne*, a game where the edges entirely determine the tiles' interiors, and there are no restrictions on the tiles' orientations. Terminology and methods will be developed in this context and then extended to handle *Galaxy Trucker* (p.45).

In *Carcassonne*, players place tiles representing terrain and structures around Carcassonne, a village in France. The game comes with 72 square tiles with three types of edges: one showing a castle, one showing a field, and one showing a road, see Figure 3.2. Players draw these tiles and must place them next to the existing tiles, matching the edges of any adjacent tiles: a castle edge must be placed against a castle edge but not against a field or a

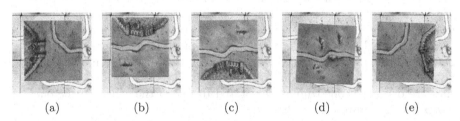

(a)   (b)   (c)   (d)   (e)

Figure 3.2: Tiles from *Carcassonne*.[2]

---

[2]Carcassonne images used by permission from Hans Im Glück. Carcassonne is copyright and trademark 2000 Hans Im Glück, and all rights are reserved worldwide.

road edge, similarly for fields and roads. Before placing the tile, players can rotate but not reflect (flip) it. Players can then claim a portion of the just-placed tile to earn victory points. As shown in Figure 3.2, tiles have the same or similar artwork, and this section will explore how many distinct tiles are possible. Of the tiles shown in Figure 3.2, Tiles 3.2b and 3.2c are equivalent in the game because one of them can be rotated to match the other. On the other hand, Tile 3.2a cannot be rotated to match any of the other tiles and therefore is distinct from the other tiles.

When a tile is placed in a particular orientation, it is a fixed tile, with the word fixed referring to the orientation (as in the case of polyominoes). Two fixed tiles are the same if they exactly match without any rotations or reflections. As fixed tiles, the tiles shown in Figure 3.2b and 3.2c are *not* the same.

Two tiles are equivalent under rotation if one can be rotated to match the other. The tiles in Figure 3.2b and 3.2c are equivalent under rotation. Two tiles are equivalent under reflection if one can be reflected or rotated to match the other. The tiles in Figure 3.2a and 3.2e are equivalent under reflection but not under rotation. When the type of equivalence is clear by context, the term equivalent will be used. If two tiles are not equivalent, they are distinct. Tiles allowed to be rotated but not reflected are often referred to as one-sided tiles, while those that can be reflected or rotated are referred to as free tiles. This matches the terminology for polyominoes in Section 2.3.

Rotation Groups

In almost all tile placement games, players can rotate a tile before placing it. Notice that most games do not specify the direction of rotation (clockwise versus counter-clockwise), as it is understood that if one is allowed to rotate a tile in one direction, one can rotate the tile in the reverse direction. Another omitted specification is the amount by which one can rotate the tile. With square tiles, the smallest rotation compatible with a square grid is 90°, but making two rotations of size 90° will result in a rotation of 180°. If a 90° rotation is allowed, it is understood that a rotation of 180° is also allowed (as would be a rotation of 270°). Given a square, how many ways can it be rotated?

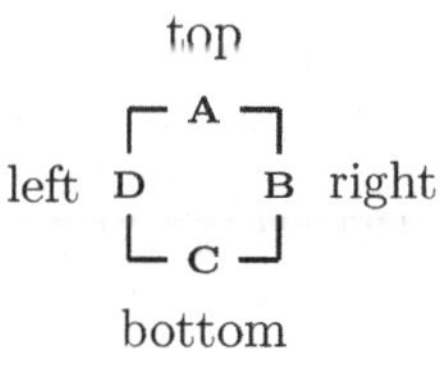

top

left D   B right

bottom

Figure 3.3: Square Labelling.

Throughout the discussion, the standard starting square tile will have edges labeled **A**, **B**, **C**, and **D** and will be oriented as shown in Figure 3.3. Notice that the boldface letters represent the edges of the tile (not the direction). After rotating it 90° clockwise, edge **A** will now be on the right, edge **B** on the bottom, edge **C** on the left, and edge **D** on the top.

The notation R_θ will be used to indicate a clockwise rotation through θ degrees. So in a tile placement game with square tiles where tiles may be

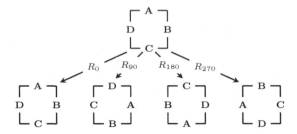

Figure 3.4: Rotations of a Square.

second	first			
	R_0	R_{90}	R_{180}	R_{270}
R_0	R_0	R_{90}	R_{180}	R_{270}
R_{90}	R_{90}	R_{180}	R_{270}	R_0
R_{180}	R_{180}	R_{270}	R_0	R_{90}
R_{270}	R_{270}	R_0	R_{90}	R_{180}

Figure 3.5: Cayley Table for Rot$_4$.

rotated (but not reflected) before being placed, the actions available would be: do not rotate the tile (R_0), rotate the tile 90° (R_{90}), rotate the tile 180° (R_{180}), and rotate the tile 270° (R_{270}). These four actions are shown in Figure 3.4. The collection of these four actions will be referred to as the rotation group of a square and denoted by Rot$_4$ (the 4 indicating the four sides of a square).

A counting argument can confirm that this includes every possible result. The key idea is that once edge **A**'s position is determined, the remaining edges are also completely determined. This is because edge **B** will need to be placed clockwise from edge **A**, edge **C** will need to be placed clockwise from edge **B**, and edge **D** will need to be placed clockwise from edge **C**. This means only four options are available, so at most, four actions can be available.

An important aspect of actions is the ability to chain them together, doing one action and then a second action. For example, a rotation of 90° could be followed by a rotation of 180°. It is often easiest to describe how the actions chain together by presenting a Cayley Table.[3] These tables generalize the multiplication tables used in primary school for students learning basic multiplication facts. Like ordinary multiplication tables, the first row and first column list the possible actions. An entry in the table shows the result of performing the action determined by the column heading followed by the action determined by the row heading. The table for the rotations of a square is given in Figure 3.5. For example, the result of R_{90} followed by R_{270} is

---

[3]These tables are named after Arthur Cayley, a nineteenth-century British mathematician who proved that all groups can be interpreted as a group of actions.

shaded in Figure 3.5. This composition is written as $R_{270}R_{90}$, where the action applied first is to the right and the action applied second is to the left. This convention of reading a product "right-to-left" is common in mathematics when actions are composed (and matches the convention when composing functions or multiplying matrices). Another important action is being able to reverse each action. For example, the action that reverses R_{90} is R_{270} because when one does these two actions one after the other (in either order), one returns the tile to its original orientation.

Suppose a collection of actions satisfies the three properties listed as Group Properties. In that case, it is referred to as a group.[4] Readers familiar with groups may have noticed that the properties omit the associative property. All groups presented in this chapter are groups of actions, and associativity is automatically satisfied.

Group Properties: A collection of actions satisfying these three properties is a group.

Identity: The action that leaves the tile in its original orientation is in the collection.

Inverse: The reverse of every action in the collection is also in the collection.

Closure: For every pair of actions in the collection, their composition (meaning to do one and then the other) is also in the collection.

Given a group of actions, two tiles are equivalent under this group if a group action changes one into the other. The first group property, the identity property, formalizes the intuitive idea that any tile should be equivalent to itself as the identity action transforms a tile to itself. The second property, the inverse property, formalizes the intuitive idea that two equivalent tiles do not require them to be referred to in any particular order, as if one tile is equivalent to a second tile through a group action, the inverse action will show that the second tile is equivalent to the first. The third property, the closure property, formalizes the intuitive idea that if one tile is equivalent to a second tile, and the second tile is equivalent to a third tile, the first and third tiles are equivalent, as the composition of the action relating the first and second tiles and the action relating the second and third tiles will show that the first and third tiles are equivalent.

---

[4]The term "group" is a technical term in mathematics and was chosen by Évariste Galois, a nineteenth-century French mathematician who used abstract algebra to prove that there is no quintic formula.

Because of this, there is a connection between the three group properties and equivalence relations. A relation is a property for pairs of objects. Common examples of relations are equality (two elements are equal to each other, $2^2 = 4$) or less-than (one element is less than another element, $3 < 4$). An equivalence relation is a relation that satisfies the three properties listed as Equivalence Relation Properties.

Equivalence Relation Properties: An equivalence relation is a relation, "equivalent to," satisfying these three properties.

Reflexivity: Every object is equivalent to itself.

Symmetry: If one object is equivalent to a second object, the second object is equivalent to the first object.

Transitivity: If one object is equivalent to a second object, and the second object is equivalent to a third object, then the first object is equivalent to the third object.

Equality is one example of an equivalence relation. On the other hand, the relation of being less than is not an equivalence relation as it does not satisfy reflexivity or symmetry. When mathematicians refer to objects as equivalent, as was done above for tiles that could be rotated to match one another, the relation between the objects is always an equivalence relation.

Returning to the actions that form Rot_4, the three group properties must be checked to verify that Rot_4 is a group. The operation R_0 leaves the tile's orientation unchanged, so Rot_4 satisfies the identity property. Each action can be undone: R_0 and R_{180} both undo themselves while R_{90} and R_{270} undo each other. So Rot_4 satisfies the inverse property. The easiest way to check that every composition of actions in Rot_4 is in Rot_4 is by looking at Figure 3.5, which shows every composition of actions, and noting that every composition results in an action that is in Rot_4. So Rot_4 satisfies the closure property.

As a counterexample, the three actions R_0, R_{90} and R_{270} do not form a group. This collection satisfies the identity property and the inverse property. Still, it is not a group because it does not satisfy the closure property: R_{90} and R_{90} are in the collection, but $R_{180} = R_{90}R_{90}$ is not in the collection.

The groups of rotations of other regular polygons are also action groups. For tabletop games, the most relevant are the rotation groups for regular triangles, Rot_3, squares, Rot_4, and regular hexagons, Rot_6, which are given below.

$$\mathrm{Rot}_3 = \{R_0, R_{120}, R_{240}\}$$
$$\mathrm{Rot}_4 = \{R_0, R_{90}, R_{180}, R_{270}\}$$
$$\mathrm{Rot}_6 = \{R_0, R_{60}, R_{120}, R_{180}, R_{240}, R_{300}\}$$

Figure 3.6: "Tiles" in *Railroad Ink*.[5]

Dihedral Groups

In some games, a player can reflect (flip) a tile before placing it. The ability to reflect a tile is common in polyomino games (discussed in Section 2.3) and roll-and-write games. In the roll-and-write game *Railroad Ink*, players strive to fill a square 7×7 grid with roadways and train tracks. To do this, one of the players rolls four dice, each showing a configuration of roads and tracks. Players can rotate or reflect each configuration and then write it into one grid square, as shown in Figure 3.6. Given a square whose edges may be blank or contain a road or a track, how many ways can it be oriented?

A square tile can be reflected across the horizontal, vertical, or one of the two diagonal axes. The notation F_θ indicates a reflection across an axis at θ degrees counterclockwise from the horizontal axis, where $0 \leq \theta < 180$. So, a square can be reflected across the horizontal axis (F_0), the vertical axis (F_{90}), and the two diagonal axes (F_{45} and F_{135}). The result of these reflections and the four rotations are found in Figure 3.7 with the initial orientation in the center.

The fact that these are the only eight orientations when allowing rotations and reflections can be confirmed with the following counting argument. Starting with the tile in Figure 3.3, edge **A** can be placed in any of the four directions. Because the tile can be reflected, edge **B** can be placed either clockwise or counter-clockwise from edge **A** (two options). Once the locations for edges **A** and **B** are determined, the remaining edges are completely determined, so the Product Rule calculates that there are $4 \times 2 = 8$ possible orientations.

---

[5]Railroad Ink™ is a game by Hjalmar Hach and Lorenzo Silva, illustrated by Marta Tranquilli, edited by Horrible Guild. ©2019-2023, Railroad Ink, Horrible Guild and their logos are trademarks of Horrible Games S.r.l.

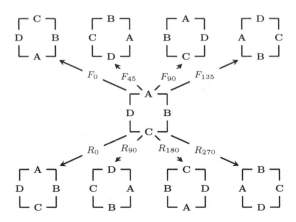

Figure 3.7: Symmetries of a Square.

Again, one could verify that these eight actions form a group by checking that the collection satisfies the three properties of a group (they do). This group is known as the dihedral group for a square, and the notation Dih_n is used for the dihedral group for a regular polygon with n sides. Similar construction for regular triangles, Dih_3, and regular hexagons, Dih_6, complete the list of dihedral groups most relevant to tabletop games.

$$\text{Dih}_3 = \{R_0, R_{120}, R_{240}, F_{30}, F_{90}, F_{150}\}$$
$$\text{Dih}_4 = \{R_0, R_{90}, R_{180}, R_{270}, F_0, F_{45}, F_{90}, F_{135}\}$$
$$\text{Dih}_6 = \{R_0, R_{60}, R_{120}, R_{180}, R_{240}, R_{300}, F_0, F_{30}, F_{60}, F_{90}, F_{120}, F_{150}\}.$$

Symmetric Groups

The symmetric group allows any side to be moved to any other side. Not all of these actions would leave a physical cardboard tile intact, but the action of this group can be applied to many game components, like shuffling a deck of cards, where any card can be placed into any position in the deck through shuffling. The symmetric group on n objects, denoted by Sym_n, consists of all permutations of those n objects. Permutations were discussed in Chapter 1, where it was noted that there are $n!$ permutations on n objects, so the size of Sym_n is $n!$. Even though Sym_3 is manageable with only six elements, Sym_4 is fairly large with 24 elements and Sym_5 has 120 elements. Rapid growth is expected as the number of elements in the symmetric group grows at a factorial rate compared to the number of objects permuted.

3.2 THE LEMMA THAT IS NOT BURNSIDE'S

In most cases from tabletop games, it is possible to determine the number of distinct tiles using brute force. However, a strangely named lemma can make the process easier and can be applied to cases with more possibilities. A lemma is usually an auxiliary proposition that will be used to prove a more significant result. This lemma was originally stated by Cauchy[6] and proven by Frobenius.[7] However, Burnside[8] published it in *Theory of Groups of Finite Order* [14] and the result became known as "Burnside's Lemma." It is now sometimes called "The Lemma that is not Burnside's" in recognition that the original name was a misattribution. Here it is attributed to Cauchy and Frobenius.

Cauchy-Frobenius Lemma: The number of distinct objects in a set is the average of inv(α) where α varies over all group elements.

To use the theorem, one needs to know what inv(α) means. The number inv(α) is the number of objects that do not change when α acts on them. It is the number of tiles that are invariant under α. To make this easier to follow, focus on the case of *Carcassonne* (p.46). In this case, the tiles are square with three types of edges (castle, field, or road), and the group will be Rot$_4$ since tiles cannot be reflected in the game.

If the tile shown in Figure 3.2d is rotated $180°$, it can still be placed in any location the original tile could be placed. So this tile is invariant under R_{180}. In this case, one could identify that the tile has been rotated by tracking the bushes and cows in the artwork, but since these have no gameplay implications, they are ignored here. However, if it is rotated by $90°$, it can no longer be placed in those locations. So, this tile is not invariant under R_{90}.

One way to determine inv(α) would be to list all the fixed tiles in the game and look at the result of applying each group action to each fixed tile. However, that would require considering $3^4 - 81$ fixed tiles under the four group actions for a total of $81 \times 4 = 324$ cases. So, consider each action and look at any restrictions on invariant tiles under that action.

R_0: The action R_0 does not change the fixed tile, so every fixed tile is invariant under this action. There are no restrictions, so each of the edges can be of any type (four selections, each with three options), and inv(R_0) is the total number of fixed tiles, which can be calculated with the Power Rule: inv(R_0) = $3^4 = 81$.

R_{90} and R_{270}: Assume a fixed tile with standard labeling (shown in Figure 3.3) is invariant under R_{90}. It will have one of three types on edge **A** (either a castle, a field, or a road). For discussion purposes, let's assume that edge **A** is a castle. Because the tile is invariant under a $90°$ rotation

---

[6] Baron Augustin-Louis Cauchy was an eighteenth–nineteenth-century French mathematician who made significant progress in complex analysis and permutation groups.

[7] Ferninand Georg Frobenius was an eighteenth–nineteenth-century German mathematician who made significant advances in the theory of differential equations and group theory.

[8] William Burnside was a nineteenth–twentieth-century English mathematician who made significant contributions to the theory of finite groups.

clockwise, edge **D** must also be a castle. This is because the rotated tile must still have a castle on its top side after rotation (or it would not be invariant). A similar argument shows that edge **C** must match edge **D** and edge **B** must match edge **C**. So, every side must have the same type. Therefore, there are only three fixed tiles that are invariant (one choice with three options) and $\text{inv}(R_{90}) = 3$. The same argument concludes that three tiles are invariant under R_{270}.

R_{180}: Assume a fixed tile with standard labeling (shown in Figure 3.3) is invariant under R_{180}. It will have one of three types on edge **A**. Because it is invariant under a rotation of 180°, edge **C** must match edge **A**. Similarly, there are three types for edge **B**, but once edge **B**'s type has been determined, edge **D** must match. Therefore, there are two choices (edge **A** and edge **B**) each with three options and the Power Rule calculates $\text{inv}(R_{180}) = 3^2 = 9$.

The average of these four values gives 24 distinct tiles:

$$\frac{1}{4}(81 + 3 + 3 + 9) = 24.$$

This can be confirmed by directly listing these 24 possibilities or grouping the 81 total tiles into piles of equivalent tiles. However, calculating how many there are before undertaking that process may determine whether it is feasible and help confirm that all possibilities have been listed.

One important aspect of the Cauchy-Frobenius Lemma is that it requires the set of actions to form a group. The set $\{R_0, R_{90}, R_{270}\}$ fails to be a group, and if one attempts to use the formula from the Cauchy-Frobenius Lemma using this set, it gives the number 29:

$$\frac{1}{3}(81 + 3 + 3) = 29.$$

However, this value does not count anything related to the problem.

This calculation can be generalized to square tiles with one of n types of edges. The calculation is similar to the case for three types of edges, but results in the quartic polynomial

$$\frac{1}{4}\left(n^4 + n + n + n^2\right).$$

While the number of tiles will grow, counting the number of tiles using a computer is still reasonable since the number of tiles exhibits polynomial growth. (As an interesting side note, this also proves that the expression $n^4 + n^2 + 2n$ is divisible by four for all positive integers n.)

In *Railroad Ink* (p.51), each edge may be empty, a road or a track. Unlike *Carcassonne*, the configuration may be reflected before being drawn onto the grid. How many distinct tiles are possible in the game *Railroad Ink*?

Because the road and track configuration can be reflected before being drawn, the appropriate group is Dih$_4$. The process is the same as in *Carcassonne*, where the number of invariant tiles for each group action is determined. The reasoning and values for rotations are the same as in *Carcassonne*, so only the reflections are discussed here.

F_0 **and** F_{90}**:** For F_0, edges **A** and **C** must match. So, there are three selections (edges **A**, **B**, and **D**), each with three options. This results in 3^3 possible tiles. Similarly, for F_{90}, edges **B** and **D** must match, and there are 3^3 possible tiles.

F_{45} **and** F_{135}**:** For F_{45}, edges **A** and **B** must match as must edges **C** and **D**. As a result, there are two selections (edges **A** and **C**), each with three options. This results in 3^2 possible tiles. Similarly, for F_{135}, edges **A** and **D** must match and edges **B** and **C** must match, and there are 3^2 possible tiles.

Using the Cauchy-Frobenius Lemma, the total number of distinct tiles is 21:

$$\frac{1}{8}\left(3^4 + 3 + 3 + 3^2 + 3^3 + 3^3 + 3^2 + 3^2\right) = 21.$$

In *Cascadia* (p.7), players place hexagonal tiles with edges of one of five types: mountains, forests, prairies, wetlands, and rivers. In this game, edges are not required to match (so a forest can be adjacent to a prairie), but matching is incentivized by earning more victory points. The tiles are one-sided, so one can rotate them but not reflect them, so the appropriate group is Rot$_6$. How many distinct tiles are possible in the game of *Cascadia*?

Figure 3.8: Hexagon Labeling.

The calculations are similar to those for *Carcassonne*. Using the starting tile shown in Figure 3.8.

R_0**:** there are $5^6 = 15625$ possible fixed tiles by the Power Rule.

R_{60} **and** R_{300}**:** Like the R_{90} case for a square, once edge **A** is determined, this must match the habitat for every side. So, only five options are available in these cases.

R_{120} **and** R_{240}**:** Once a habitat has been determined for edge **A**, the same habitat must also be on edges **C** and **E**. Similarly, the habitat on edge **B** must also be on edges **D** and **F**. So there are $5^2 = 25$ possibilities by the Power Rule.

R_{180}**:** Finally, for a 180 degree rotation, edges **A** and **D** must match, as must edges **B** and **E**, and edges **C** and **F**. So, three edges can be freely chosen, and there are $5^3 = 125$ possibilities by the Power Rule.

The Cauchy-Frobenius Lemma computes a total of 2635 tiles:

$$\frac{1}{6}\left(5^6 + 2 \times 5 + 2 \times 5^2 + 5^3\right) = 2635.$$

Directly listing all possibilities by hand would be infeasible. Even if these options were enumerated by computer, the number of results makes such a list unwieldy. This is an unreasonably large number of tiles for a reasonably priced tabletop game, and the game *Cascadia* does not have nearly this many tiles (it has 85 tiles). Rather than attempting to represent every possible tile, the game uses precisely two types of tiles: tiles with all one habitat and tiles split into two halves with a different habitat on each half. There are two ways to calculate the number of distinct tiles in *Cascadia*.

The fastest method is a counting technique, as five tiles have precisely one type of edge, and there are $C(5,2) = 10$ tiles that will have two different types of terrain for a total of $5 + 10 = 15$ types of tiles.

Alternatively, one could use the Cauchy-Frobenius Lemma. This is facilitated by noting that if the tile has two matching edges separated by one edge, then the middle edge must also match. For example, if edges **A** and **C** are both prairies, then edge **B** must also be a prairie.

R_0: there are five tiles with the same habitat on every edge. For the tiles with two habitats, there are five choices for the habitat on edge **A**, leaving four choices for the habitat on edge **D** since opposite edges must have different habitats. The final choice is whether the habitat on edge **A** is on edges **E-F-A**, edges **F-A-B**, or edges **A-B-C** (so three choices). There are $5 \times 4 \times 3 = 60$ tiles with two habitats and $5 + 60 = 65$ total fixed tiles.

R_{60} and R_{300}: Like the R_{90} case for a square, once the habitat for edge **A** is determined, this must match the habitat for every side. So there are only five possibilities.

R_{120} and R_{240}: Once a habitat has been determined for edge **A**, the same habitat must also be on edges **C** and **E**. However, this also forces edges **B**, **D**, and **F** to share that habitat. So there are five possibilities.

R_{180}: Finally, for a 180 degree rotation, edges **A** and **D** must be the same, but this can only happen if all edges of the tile are the same (as edges **A** and **D** are on opposite sides of the tile). Again, there are only five possibilities.

Applying the Cauchy-Frobenius Lemma, *Cascadia* has 15 distinct tiles:

$$\frac{1}{6}\left(65 + 5 + 5 + 5 + 5 + 5\right) = 15.$$

Each of the tiles with precisely one type of edge is duplicated five times, and each of the tiles with two types of edges is duplicated six times, for a total of 85 tiles:

$$5 \times 5 + 6 \times 10 = 85.$$

Returning to *Galaxy Trucker* (p.45), there are four possible sides: no connector, a single-connector, a double-connector, and a triple-connector. One restriction must be considered: at least one of the tile's edges must have a connector. To account for this, each count must be reduced by one. There are also 15 interiors, ten of which are symmetric and can be placed in any orientation, and five have a "front" edge. The ten symmetric interiors are one crew cabin, two battery modules, four cargo holds, two life-support modules, and one structural component. The calculation for these tiles is the same as for *Carcassonne* (p.46), and there are 69 fixed tiles for each symmetric interior:

$$\frac{1}{4} \left(4^4 + 4^1 + 4^2 + 4^1 \right) - \underbrace{1}_{\text{unconnected}} = 69.$$

For the five that are oriented, choose a standard orientation and count the possible edges:

Cannons and Double Cannons: can be placed so the canons face any direction. To count them, standardize their orientation so the canon faces the front. These tiles will have an empty front edge, but each of the remaining three edges may have any edge type (but at least one must be a connector).

Engines and Double Engines: must be placed so the engines face the rear. To count them, standardize their orientation so the engine faces the rear. These tiles will have an empty rear edge, but each of the remaining three edges may have any edge type (but at least one must be a connector).

Shields: can be rotated so the shield protects one corner when the tile is placed. To count them, standardize their orientation so that the shield protects the front-left corner of the ship. Any edge may have any edge type (but at least one must be a connector).

The calculation for Cannons, Double Cannons, Engines, and Double Engines is the same. Three edges will be determined with four options for each edge, for a total of $4^3 - 1 = 63$ tiles (after removing the tile with no connectors). The calculation for the Shields is similar, with four edges to be determined with four options for each edge, for a total of $4^4 - 1 = 127$ tiles.

Putting this all together, there are 1069 possible tiles in *Galaxy Trucker*:

$$10 \times 69 + 4 \times 63 + 1 \times 127 = 1069.$$

The actual game consists of only 140 tiles, so many of these possibilities were not included. For example, of the 69 possible crew cabin tiles possible, only

(a) Crew Cabins.

(b) Cannon.

(c) Engine.

(d) Energy Store.

Figure 3.9: Tiles from *Galaxy Trucker*.[9]

top	right	bottom	left
Empty	Empty	Single	Triple
Empty	Empty	Triple	Double
Empty	Single	Empty	Triple
Empty	Single	Single	Triple
Empty	Single	Double	Single
Empty	Single	Double	Triple
Empty	Single	Triple	Single
Empty	Double	Empty	Triple
Empty	Double	Single	Double
Empty	Double	Triple	Single
Empty	Double	Triple	Double
Empty	Triple	Double	Double
Single	Single	Single	Double
Single	Double	Single	Double
Single	Double	Single	Double
Single	Double	Double	Double

Edge Type on Side:

Figure 3.10: Edges of the 16 Crew Cabin Tiles in *Galaxy Trucker*.

16 tiles are included (listed in Figure 3.9). Interestingly, a duplicated tile is included even with 53 unused possibilities.

Considering tiles with m sides and n types of edges, the Cauchy-Frobenius Lemma will produce an m degree polynomial in the variable n. So, the number of tiles will grow at a rate that is exponential in m and polynomial in n. Increasing the number of sides would be more effective if the goal is more distinct tiles. The catch is that this would require the board to be hyperbolic (see Section 2.4).

---

[9]Galaxy Trucker images used by permission from Czech Games Edition. Galaxy Trucker is ©2007 Czech Games Edition, and all rights are reserved worldwide.

3.3   THE ORBIT STABILIZER THEOREM

The previous section determined the number of distinct tiles in a collection. This section will determine the related question: how many tiles are equivalent to a single tile? For example, how many tiles are equivalent to the tile in Figure 3.2a?

These questions can be answered with the Orbit Stabilizer Theorem.

Orbit Stabilizer Theorem: The number of objects equivalent to a given object equals the number of actions divided by the number of actions that leave the object fixed.

The name of the theorem comes from two mathematical terms. The orbit of an object is the set of all objects equivalent to the given object, and the stabilizer of an object is all of the actions that leave the object fixed. Tiles have six or fewer edges for tabletop games, and the groups have no more than twelve elements. In these cases, applying all of the actions to a tile to determine its orbit is possible. Still, the theorem generalizes to situations with much larger numbers.

For example, calculating the number of tiles equivalent to the tile in Figure 3.2a from *Carcassonne* (p.46) can be done as follows. The group is Rot_4 of which only R_0 leaves the tile fixed, so there are $4/1 = 4$ equivalent tiles.

Since both the Cauchy-Frobenius Lemma and the Orbit Stabilizer Theorem are attempting to count objects under an equivalence relation, there should be some connection between the results, and there is. This connection is visible if all the objects are placed in rows, where each row corresponds to equivalent objects. Then the Cauchy-Frobenius Lemma will compute how many rows are needed, and the Orbit Stabilizer Theorem will compute how many objects are in each row.

For example, consider all square tiles whose sides can be either red, R, or green, G, under the action of Rot_4. The Power Rule indicates that there are $2^4 = 16$ total tiles. The Cauchy-Frobenius Lemma indicates that there will be six distinct tiles:

$$\frac{1}{4}\left(\text{inv}(R_0) + \text{inv}(R_{90}) + \text{inv}(R_{180}) + \text{inv}(R_{270})\right) = \frac{1}{4}(16 + 2 + 4 + 2) = 6.$$

The Orbit Stabilizer Theorem can determine the number of tiles in each row (which will differ from row to row). This organization of the 16 tiles is presented in Figure 3.11.

Counting

The Orbit Stabilizer Theorem can also be used to derive the binomial and multinomial coefficients from Chapter 1. This will be illustrated by determining the formula for $C(4, 2)$, the number of ways to select two elements

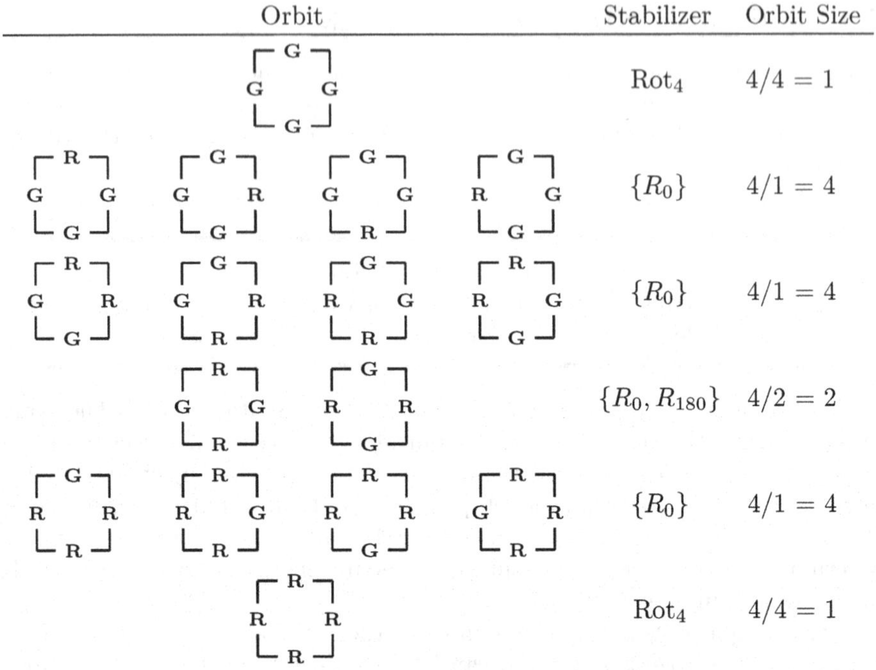

Orbit				Stabilizer	Orbit Size

Figure 3.11: Organizing Tiles by Orbits.

from a four-element set. Let $\mathcal{X}$ be the collection of all two-element subsets of $\{1, 2, 3, 4\}$, and the variable X will be used for an arbitrary element of $\mathcal{X}$. By the Combination Rule, $C(4, 2) = |\mathcal{X}|$. In this case, $\mathcal{X}$ only contains six elements and be completely listed:

$$\mathcal{X} = \{\{1, 2\}, \{1, 3\}, \{1, 4\}, \{2, 3\}, \{2, 4\}, \{3, 4\}\}.$$

The group Sym_4 acts on the set $\{1, 2, 3, 4\}$, each $\sigma \in \text{Sym}_4$ permutes the elements of $\{1, 2, 3, 4\}$. For example, assume that $\sigma \in \text{Sym}_4$ swaps the elements 1 and 3 and leaves 2 and 4 unchanged. So $\sigma(1) = 3$, $\sigma(2) = 2$, $\sigma(3) = 1$, and $\sigma(4) = 4$. Then σ will also act on $\mathcal{X}$ by replacing every 1 with a 3 and every 3 with a 1 in each set in $\mathcal{X}$.

X	$\{1, 2\}$	$\{1, 3\}$	$\{1, 4\}$	$\{2, 3\}$	$\{2, 4\}$	$\{3, 4\}$
$\downarrow$	$\downarrow$	$\downarrow$	$\downarrow$	$\downarrow$	$\downarrow$	$\downarrow$
$\sigma(X)$	$\{3, 2\}$	$\{3, 1\}$	$\{3, 4\}$	$\{2, 1\}$	$\{2, 4\}$	$\{1, 4\}$

Consider $\{1, 2\} \in \mathcal{X}$. For any other $X \in \mathcal{X}$, there will be an element of Sym_4 which takes $\{1, 2\}$ to X by taking 1 to the smaller element of X and 2 to the larger element of X. So the orbit of $\{1, 2\}$ will be all of $\mathcal{X}$. Each element of Sym_4 which leaves $\{1, 2\}$ invariant can permute the set $\{1, 2\}$ and can permute the set $\{3,4\}$ but it cannot take an element from one of these sets to the other set. There are 2! ways to permute $\{1,2\}$ and 2! ways to permute $\{3,4\}$, so the stabilizer contains $2!2! = 4$ actions. Using the notation of stab(X) for the stabilizer of X and putting this all together in the Orbit Stabilizer Theorem:

$$\binom{4}{2} = |\mathcal{X}| = \frac{|\mathrm{Sym}_4|}{|\mathrm{stab}(\{1,2\})|} = \frac{4!}{2!2!}.$$

A similar argument for larger values of n produces the general formula, where there are $k!$ ways to permute $\{1,\ldots,k\}$ and $(n-k)!$ ways to permute $\{k+1,\ldots,n\}$, so there are $k!(n-k)!$ elements in the stabilizer, resulting in the general formula

$$\binom{n}{k} = |\mathcal{X}| = \frac{|\mathrm{Sym}_n|}{|\mathrm{stab}(\{1,2,\ldots,k\})|} = \frac{n!}{k!(n-k)!}.$$

And by considering more than one subset of $\mathcal{X}$, the argument can also be made to apply to the multinomial coefficients.

This last section has been somewhat abstract, so it is time to return to more direct applications to tabletop games. However, it is interesting to note that many of the geometric problems that involve symmetry are solvable by techniques in group theory, a sub-discipline of algebra.

CHAPTER 4

Graph Theory

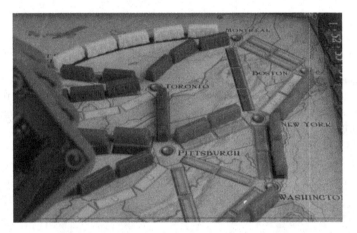

Figure 4.1: Components from *Ticket To Ride*.[1]

In *Ticket To Ride*, players attempt to build a railroad network across North America by claiming routes between cities on the game board. Players discard sets of cards to claim routes and earn points. In Figure 4.1, the player can discard three blue cards to claim the route from Montréal to New York.

What are the shortest routes from Montréal to Los Angeles, and how much do they score?

---

[1] Images of Ticket to Ride used with permission. Days of Wonder, the Days of Wonder logo, and Ticket to Ride are all trademarks or registered trademarks of Days of Wonder, Inc. All Rights Reserved.

     DOI: 10.1201/9781003383529-4

INTRODUCTION

Movement on a grid is common in games like *Undaunted: Normandy* and *Hoplomachus: Remastered*, which simulate small-scale skirmishes. However, a graph (or network) is more appropriate in games like *Fury of Dracula* and *Ticket To Ride*, which simulate movement between cities. Graphs will also be used to visualize relations between rewards in a game or between situations in gameplay.

The analysis of graphs here is primarily done with matrices, which are introduced in this chapter. Once the information about the graph has been encoded into a matrix, a computer algebra system can do the bulk of the required computations to determine information about the graph. In particular, the multiplication of matrices can solve questions involving the distance between locations or the number of routes between locations.

More careful information encoding using generating functions (from Chapter 1) allows matrices to track other route features between two locations. For example, in the case of *Ticket To Ride*, matrices can track the required cards, the victory points earned, and the cities visited on each route.

4.1 ADJACENCY IN GRAPHS

As before, some more straightforward questions will be answered before returning to the opening question. In *Fury of Dracula*, one player takes the role of Dracula while the other players take the role of hunters tracking Dracula through Europe. The graph for land travel in the Iberia region is shown in Figure 4.2. Madrid is the unlabeled central city in both graphs. The left graph shows which towns and cities are connected by roads, and the right graph shows which towns and cities are connected by railroads. How many three- or four-leg routes from Granada to Lisbon are in the Iberia region?

A graph consists of points (called vertices) together with edges that connect pairs of vertices. For example, the vertices and edges of the tilings and dual tilings studied in Chapter 2 are graphs. While the tilings from Chapter 2 have an infinite number of vertices and edges in the case of the plane or the

(a) Road Travel. (b) Rail Travel.

Figure 4.2: Travel Options in *Fury of Dracula*.

hyperbolic plane, the graphs in this chapter will have a finite number of vertices and edges. Many of these graphs represent geographical locations, where the vertices are places of interest, and the edges show one step of movement.

Several terms used in graph theory will be introduced using the graphs shown in Figure 4.2. There is some variation in terminology, but the terminology used in this book is common. Two vertices are adjacent if there is an edge connecting them. So, Madrid (in the center) and Granada (lower right) are adjacent to each other in the road graph but are not adjacent to each other in the rail graph. The term incident is used to indicate that a vertex is the endpoint of an edge. For example, Santander is incident to the edge between Santander and Saragossa. Also, the edge between Santander and Saragossa is incident to the vertex Santander. The number of edges incident to a vertex is the degree of that vertex. Madrid has a degree of six in the road graph, and Granada has a degree of zero in the rail graph. A vertex with a degree of zero is an isolated vertex, so Granada is an isolated vertex in the rail graph. A vertex with a degree of one is a leaf, so Barcelona is a leaf in the road graph.

A walk through the graph is a sequence of adjacent vertices. A walk can repeat vertices and edges multiple times. In the road graph, one walk would be Saragossa–Santander–Madrid–Saragossa–Barcelona. If a graph has multiple edges between the same vertices, specifying which edge is being used may be required, but in many cases, there will only be one edge, or the choice of the edge will be unimportant. Each edge in a walk is referred to as a hop, so the walk above is a four-hop walk from Saragossa to Barcelona. A path is a walk that never visits any vertex more than once except the first and last vertices, which may be the same vertex. In the case of a path where the first and last vertices are the same, the path is a cycle. The example of the walk above is not a path, as it visits Saragossa twice. The walk Saragossa–Santander–Madrid–Saragossa is a cycle. A graph is connected if there is a path from every vertex to every other vertex. The road graph is connected, while the rail graph is not.

For a small graph, finding the shortest route is not too difficult, but the entire graph of *Fury of Dracula* has 70 vertices and 396 edges, so hand-counting long walks becomes tedious and error-prone. Matrices will be introduced to provide a technique to answer the question above in a way that can be generalized to larger graphs.

Matrices

Matrices are mathematical tools that can be used to analyze graphs, among other things. A matrix is a rectangular array of numbers often surrounded by a set of parentheses to delineate it. The dimension of the matrix is given by the number of rows, n, followed by the number of columns, m, and is often written as $n \times m$ and read as "n by m." The element in row i and column j is referred to as the (i, j) entry. The (i, j) entry of the matrix $\mathbf{M}$ is denoted by $\mathbf{M}[i, j]$.

$$\begin{pmatrix} 1 & 2 \\ 3 & 4 \end{pmatrix} \qquad \begin{pmatrix} 1 & 1 & 1 & 1 \end{pmatrix} \qquad \begin{pmatrix} 1 & 0 & 0 \\ 0 & 1 & 0 \\ 0 & 0 & 1 \end{pmatrix} \qquad \begin{pmatrix} 1 & 2 \\ 3 & 4 \\ 5 & 6 \end{pmatrix}$$

(a) 2×2 (b) 1×4 (c) 3×3 identity   (d) 3×2

Figure 4.3: Matrices.

Some matrices of various dimensions are shown in Figure 4.3. A matrix is a square matrix if it has the same number of rows as it has columns. Matrices in Figures 4.3a and 4.3c are square matrices. A square matrix with ones down the main diagonal and zeros in every other entry is known as an identity matrix. There is an $n \times n$ identity matrix for every n, and they will be denoted by $\mathbf{I}$ with the dimension determined by context. The matrix in Figure 4.3c is an identity matrix. A matrix is a row matrix if it has only one row and a column matrix if it has only one column. The matrix in Figure 4.3b is a row matrix. A matrix of all of whose entries are one will be denoted by $\mathbf{1}$ with the dimension determined by context. This is *not* the same as the identity matrix. The matrix in Figure 4.3b would be denoted by $\mathbf{1}$.

One advantage of matrices is that each entry corresponds to a *row* and a *column*, so they can represent tabular information. Here, they will be used to represent connections between the vertices of a graph. The row index will indicate the vertex *to which* one is moving, and the column will indicate the vertex *from which* one is moving. A matrix constructed in this way is an adjacency matrix.

To construct an adjacency matrix, first decide on an order for the vertices. This choice is arbitrary but must be used consistently. For the cities in the Iberian region shown in Figure 4.2, the order used will be English reading order: left to right and then down. Some examples will label the rows and columns in gray with the corresponding cities for clarity. These labels are not part of the mathematical notation but can be helpful, mainly when multiple different orderings are possible.

The adjacency matrix for traveling by road is determined as follows. In an adjacency matrix, the (i, j) entry is the number of edges from vertex j to vertex i. In *Fury of Dracula*, there is at most one road between any two cities. So, values in this adjacency matrix are either zero, meaning the two locations are not adjacent or one, meaning the two locations are adjacent. Another feature of *Fury of Dracula* that makes this calculation easier than the general case is that all edges are bidirectional (two-way), so the (j, i) entry will equal the (i, j) entry.

So, the value in the first column and first row is zero because there is no edge between Santander and itself. The value in the first column and second row is one because there is an edge between Santander (the starting point) and

	Sant	Sara	Bar	Lis	Mad	Ali	Cad	Gra	deg
Sant	0	1	0	1	1	0	0	0	3
Sara	1	0	1	0	1	1	0	0	4
Bar	0	1	0	0	0	0	0	0	1
Lis	1	0	0	0	1	0	1	0	3
Mad	1	1	0	1	0	1	1	1	6
Ali	0	1	0	0	1	0	0	1	3
Cad	0	0	0	1	1	0	0	1	3
Gra	0	0	0	0	1	1	1	0	3
deg	3	4	1	3	6	3	3	3	

Figure 4.4: Adjacency Matrix for Road Travel in *Fury of Dracula*.

	Sant	Sara	Bar	Lis	Mad	Ali	Cad	Gra	deg
Sant	0	0	0	0	1	0	0	0	1
Sara	0	0	1	0	1	0	0	0	2
Bar	0	1	0	0	0	1	0	0	2
Lis	0	0	0	0	1	0	0	0	1
Mad	1	1	0	1	0	1	0	0	4
Ali	0	0	1	0	1	0	0	0	2
Cad	0	0	0	0	0	0	0	0	0
Gra	0	0	0	0	0	0	0	0	0
deg	1	2	2	1	4	2	0	0	

Figure 4.5: Adjacency Matrix for Rail Travel in *Fury of Dracula*.

Saragossa (the ending point). The value in the third row will be zero because there is no road from Santander to Barcelona. Filling out this matrix results in the matrix shown in Figure 4.4. An extra row and column have also been added outside the matrix labeled with the degree of each vertex (obtained by summing across the rows and down columns). Like the city names, this is not part of the matrix but can be helpful when checking the matrix for accuracy. Figure 4.5 gives the matrix for travel by railroad.

In *Fury of Dracula*, a hunter character can travel by rail or road. To sum two matrices, one adds the corresponding entries. The adjacency matrix for either taking one hop by road or one hop by rail would be the sum of the matrix for road travel and the matrix for rail travel. This is done in Figure 4.6 for road and rail travel in *Fury of Dracula*. This is consistent with the Sum Rule from Chapter 1, where a sum is used when one can select from more than one collection (here, the sum represents traveling using either a road edge or a rail edge).

$$\mathbf{A} = \begin{pmatrix} 0 & 1 & 0 & 1 & 2 & 0 & 0 & 0 \\ 1 & 0 & 2 & 0 & 2 & 1 & 0 & 0 \\ 0 & 2 & 0 & 0 & 0 & 1 & 0 & 0 \\ 1 & 0 & 0 & 0 & 2 & 0 & 1 & 0 \\ 2 & 2 & 0 & 2 & 0 & 2 & 1 & 1 \\ 0 & 1 & 1 & 0 & 2 & 0 & 0 & 1 \\ 0 & 0 & 0 & 1 & 1 & 0 & 0 & 1 \\ 0 & 0 & 0 & 0 & 1 & 1 & 1 & 0 \end{pmatrix}$$

Figure 4.6: Adjacency Matrix for Iberian Travel in *Fury of Dracula*.

Matrix Multiplication

While matrix addition can be useful, what makes matrices fantastic is that they can be multiplied together, and these products convey information about walks with more than one hop. When defining the product of two adjacency matrices, it should be done to match the interpretation of products in the Product Rule, namely, making consecutive choices, one for each factor. Here, the product calculates the number of walks that can be taken by traveling along a walk counted in the matrix on the right, followed by a walk counted in the matrix on the left. To demonstrate how matrices can be used to track possible paths, consider the simplified version of the *Fury of Dracula* graph shown in Figure 4.7, where some edges have been removed. How many two-hop routes from Santander to Cadiz pass through Lisbon, Madrid, or Alicante?

Edges in the figure indicate available routes (either road or railroad). Notice that in this graph, some vertices have multiple edges connecting them. In particular, a hunter can travel from Madrid to Santander by either road or

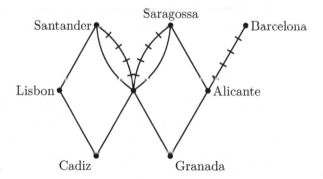

Figure 4.7: Simplified Travel Options in *Fury of Dracula*.

rail. The first leg of the journey has the adjacency matrix

$$
\begin{array}{c}
 & \begin{array}{ccc} Sant & Sara & Bar \end{array} \\
\begin{array}{c} Lis \\ Mad \\ Ali \end{array} & \left(\begin{array}{ccc} 1 & 0 & 0 \\ 2 & 2 & 0 \\ 0 & 1 & 1 \end{array} \right).
\end{array}
$$

This matrix restricts the starting location to Santander, Saragossa, or Barcelona and the destination to Lisbon, Madrid, or Alicante.

The second leg of the journey uses the adjacency matrix

$$
\begin{array}{c}
 & \begin{array}{ccc} Lis & Mad & Ali \end{array} \\
\begin{array}{c} Cad \\ Gra \end{array} & \left(\begin{array}{ccc} 1 & 1 & 0 \\ 0 & 1 & 1 \end{array} \right).
\end{array}
$$

This matrix restricts the starting location to Lisbon, Madrid, or Alicante and the destination to Cadiz or Granada.

Matrices are read right-to-left, so the trip's first leg is the matrix on the right, and the trip's second leg will be the matrix on the left. This convention is applied in mathematics for functions and matches the convention from Chapter 3, where actions were read from right to left as group actions transformed objects. Here, the matrices transform the hunter's potential locations. Therefore, the number of routes from each northern city to each southern city is given in the product of the matrices,

$$
\begin{array}{c}
 & \begin{array}{ccc} Lis & Mad & Ali \end{array} \\
\begin{array}{c} Cad \\ Gre \end{array} & \left(\begin{array}{ccc} 1 & 1 & 0 \\ 0 & 1 & 1 \end{array} \right)
\end{array}
\begin{array}{c}
 & \begin{array}{ccc} Sant & Sara & Bar \end{array} \\
\begin{array}{c} Lis \\ Mad \\ Ali \end{array} & \left(\begin{array}{ccc} 1 & 0 & 0 \\ 2 & 2 & 0 \\ 0 & 1 & 1 \end{array} \right)
\end{array}
$$

$$
= \begin{array}{c}
 & \begin{array}{ccc} Sant & Sara & Bar \end{array} \\
\begin{array}{c} Cad \\ Gra \end{array} & \left(\begin{array}{ccc} ? & ? & ? \\ ? & ? & ? \end{array} \right).
\end{array}
$$

The product's dimension will be 2×3 because the matrix counts the number of routes from the three northern cities to the two southern cities. Focusing on traveling from Santander to Cadiz, one must start in Santander. The Santander column of the right matrix is shaded to indicate that this is the starting location. The walk will end in Cadiz, and the Cadiz row of the left matrix has been shaded to indicate that this is the ending location. The number of routes will be placed in the shaded cell in the product. To count the number of routes from Santander to Cadiz, collect routes that pass through each intermediate point: Lisbon, Madrid, or Alicante. If traveling through Lisbon, there is one way to travel from Santander to Lisbon (in the right matrix) followed by one way to travel from Lisbon to Cadiz (in the left matrix), for a total of $1 \times 1 = 1$ route. If traveling through Madrid, there are two ways to travel from Santander to Madrid (in the right matrix) followed by one way to travel from Madrid to Cadiz (in the left matrix), for a total of $2 \times 1 = 2$ routes. If traveling through Alicante, there are zero ways to travel from Santander to Alicante (in the right matrix), followed by zero ways to travel from Alicante to Cadiz (in the left matrix), for a total of $0 \times 0 = 0$ routes. Adding all the possible routes gives $1 + 2 + 0 = 3$ total routes from Santander to Cadiz.

A similar calculation can be done for the remaining two-hop routes between northern and southern cities. Each new entry will be a sum of three terms, where each term represents travel through one of the intermediate cities. The computed product is

$$
\begin{array}{c}
 \\
Cad \\
Gre
\end{array}
\begin{pmatrix}
Lis & Mad & Ali \\
1 & 1 & 0 \\
0 & 1 & 1
\end{pmatrix}
\begin{array}{c}
 \\
Lis \\
Mad \\
Ali
\end{array}
\begin{pmatrix}
Sant & Sara & Bar \\
1 & 0 & 0 \\
2 & 2 & 0 \\
0 & 1 & 1
\end{pmatrix}
$$

$$
= \begin{array}{c}
 \\
Cad \\
Gra
\end{array}
\begin{pmatrix}
Sant & Sara & Bar \\
1+2+0 & 0+2+0 & 0+0+0 \\
0+2+0 & 0+2+1 & 0+0+1
\end{pmatrix}
= \begin{pmatrix}
3 & 2 & 0 \\
2 & 3 & 1
\end{pmatrix}.
$$

From here forward, the *interpretation* of matrix multiplication as taking successive walks through a graph is important, not the actual mechanics of computing the product, and the actual computation will be relegated to a computer algebra system. However, if one ever needs to multiply two matrices, one can use the interpretation to guide one's computation.

The square of an adjacency matrix, $\mathbf{A}^2$, will describe where a hunter can move in two hops, and its cube, $\mathbf{A}^3$, will describe where a hunter can move in three hops. In general, its nth power, $\mathbf{A}^n$, will describe where a hunter can move in n hops. The identity matrix, $\mathbf{I}$, describes where one can move in zero hops, consistent with the convention that anything to the power zero is the identity. These matrices track walks, so they typically overcount the number of *paths* (which do not revisit a vertex except possibly the first). This overcounting is irrelevant when the issue is whether one can reach the location (the value is positive) or not (the value is zero). There will be no overcounting when counting the shortest routes between two locations.

Returning to how many three- or four-hop walks go from Granada to Lisbon, raising the matrix in Figure 4.6, $\mathbf{A}$, to the third power results in the matrix in Figure 4.8. Starting in Granada corresponds to the last column, and ending in Lisbon corresponds to the fourth row. The shaded cell, $\mathbf{A}^4[4, 8] = 9$, shows there are nine possible three-hop walks. This can be verified with some patience.

There are 69 four-hop walks from Granada to Lisbon, which can be found as the $(4, 8)$ entry of the fourth power of the adjacency matrix: $\mathbf{A}^4[4, 8] =$

$$\mathbf{A}^3 = \begin{pmatrix} 16 & 23 & 13 & 17 & 43 & 16 & 10 & 12 \\ 23 & 20 & 26 & 14 & 55 & 22 & 12 & 12 \\ 13 & 26 & 4 & 14 & 11 & 19 & 7 & 8 \\ 17 & 14 & 14 & 12 & 43 & 14 & 12 & 9 \\ 43 & 55 & 11 & 43 & 34 & 49 & 24 & 24 \\ 16 & 22 & 19 & 14 & 49 & 16 & 9 & 13 \\ 10 & 12 & 7 & 12 & 24 & 9 & 6 & 9 \\ 12 & 12 & 8 & 9 & 24 & 13 & 9 & 6 \end{pmatrix}$$

Figure 4.8: Matrix for Three-Hops in Iberia in *Fury of Dracula*.

69. At first, this seems very high, but it does consider all walks, not just paths. This means routes such as Granada–Madrid–Lisbon–Madrid–Lisbon are included. Section 4.3 discusses counting paths.

For the complete game of *Fury of Dracula*, there are 60 cities, ten spaces representing ship travel, and two railway systems (a slower yellow system and a faster white system). In one turn of the game, a hunter may stay at their location, move one hop on the roads, one or two hops on either railroad system, move three hops on the white railroad system, or move one hop on the shipping system. This results in five adjacency matrices.

I: The identity matrix represents the player's choice not to move from their current location.

R: The adjacency matrix for the road graph, where players can choose to move one hop.

Y: The adjacency matrix for the yellow railroad, where players can choose to move one or two hops.

W: The adjacency matrix for the white railroad, where players can choose to move one, two, or three hops.

S: The adjacency matrix for the shipping system, where players can choose to move one hop.

The possible moves can be computed with the 70×70 adjacency matrix

$$\mathbf{I} + \mathbf{R} + (\mathbf{Y} + \mathbf{W}) + (\mathbf{Y} + \mathbf{W})^2 + \mathbf{W}^3 + \mathbf{S}.$$

When Dracula is revealed on the board, hunters will want to travel to his location. For a hunter not on another mission, being at a location that allows easy access to other cities may be beneficial. By computing the powers of the adjacency matrix in a computer algebra system, one can determine the number of cities within a specified number of turns from a given city. For example, the number of cities that can be reached from Paris in one turn is the number of non-zero entries in the column associated with Paris. Similarly,

Table 4.1: Victory Points in *Ticket To Ride*.

Route Length	Victory Points
1	1
2	2
3	4
4	7
5	10
6	15

for the other 69 spaces. When this is done, it is discovered that Paris and Cologne have the most cities within one turn's movement, each with 16.

Powers of the matrix indicate that 18 cities can reach all 60 cities within five turns. Raising the adjacency matrix to higher powers shows no walks taking seven or fewer turns between Athens and Manchester or between Dublin and any of Castle Dracula, Constanta, Galatz, Sofia, Valona, and Varna. Raising the matrix to the eighth power shows that all cities can be reached from all other cities in eight or fewer turns.

4.2 MATRICES OF GENERATING FUNCTIONS

In *Ticket To Ride* (p.62), routes are claimed by discarding a set of cards with the same color as the route and equal to the route length to claim a route. Players then score between one and 15 points based on the route length between the two cities, as shown in Table 4.1. For example, the route between Montréal and New York has three blue spaces, requires three blue cards to claim and will score a player four points. While the adjacency of cities is important, so are the number and type of cards required to claim routes. The generating functions from Section 1.4 can incorporate this information into adjacency matrices. Consider the very small version of the *Ticket To Ride* board shown in Figure 4.9. The letters on the edges indicate the cards required to claim that route. To claim a route from Boston to New York would require one of the two color combinations: two yellows or two reds. Wild cards present a slight issue since a route from Montréal to Pittsburgh through Toronto would require four cards forming two pairs (for example, two red and two blue, but not three red and one blue). This section will ignore this aspect of gameplay, and wildcards will be treated as another color. Similarly, the focus will be on the number of cards required rather than how they are matched. A graph where each edge has a value (weight) associated with it, as in Figure 4.9, will be referred to as a weighted graph. This graph shows that it costs two cards (of the same color) to claim the route from Montréal to Toronto and three blue cards to claim the route from Montréal to New York. One way to encode these weights into the adjacency matrix is to fill the matrix with values that encode information about each edge. Such a matrix is called a weighted adjacency

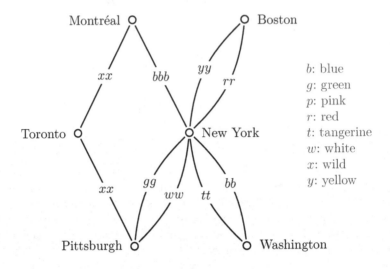

Figure 4.9: Simplified Map for *Ticket To Ride*.

matrix or sometimes a transition matrix. However, this matrix must meet the following criteria to be useful. It must allow multiple edges between the same vertices with different costs (for example, the two edges from Boston to New York represent the cost of two yellow or two red cards), and chaining together routes should add the costs when the matrices are multiplied.

When multiplying two values with the same base, the exponents are added: $x^n x^m = x^{n+m}$. This technique was presented in Section 1.4 to count the possible opening hands in *Brass: Birmingham* (p.1). The total number of cards used can be tracked by choosing a variable, say x, and placing the number of required cards as the exponent in that variable. The route from Montréal to New York requires three blue cards and would be represented by x^3, and the route from New York to Washington requires either two tangerine cards or two blue cards and would be represented by $x^2 + x^2 = 2x^2$. Multiplying these two expressions yields $2x^5$, which indicates the cost (five in the exponent) and number (two as the coefficient) of routes from Montréal to Washington passing through New York.

Figure 4.10 shows the weighted adjacency matrices using this convention. The matrix in Figure 4.10a represents routes from Toronto and New York to Pittsburgh and Washington. The matrix in Figure 4.10b represents routes from Montréal and Boston to Toronto and New York.

The computation for travel from the northern cities (Montréal and Boston) to the southern cities (Pittsburgh and Washington) with an intermediate step

$$
\begin{array}{c}
\quad\quad Tor \quad NY \\
\begin{array}{c} Pitt \\ Wash \end{array}
\left(
\begin{array}{cc}
x^2 & 2x^2 \\
0 & 2x^2
\end{array}
\right)
\end{array}
\qquad
\begin{array}{c}
\quad\quad Mon \quad Bos \\
\begin{array}{c} Tor \\ NY \end{array}
\left(
\begin{array}{cc}
x^2 & 0 \\
x^3 & 2x^2
\end{array}
\right)
\end{array}
$$

(a) Southern Routes.   (b) Northern Routes.

Figure 4.10: Weighted Adjacency Matrices for *Ticket To Ride*

in Toronto or New York is completed by multiplying these matrices,

$$
\begin{array}{c}
\quad\quad Tor \quad NY \\
\begin{array}{c} Pitt \\ Wash \end{array}
\left(
\begin{array}{cc}
x^2 & 2x^2 \\
0 & 2x^2
\end{array}
\right)
\end{array}
\begin{array}{c}
\quad\quad Mon \quad Bos \\
\begin{array}{c} Tor \\ NY \end{array}
\left(
\begin{array}{cc}
x^2 & 0 \\
x^3 & 2x^2
\end{array}
\right)
\end{array}
$$

$$
=
\begin{array}{c}
\quad\quad\quad Mon \quad\quad Bos \\
\begin{array}{c} Pitt \\ Wash \end{array}
\left(
\begin{array}{cc}
x^4 + 2x^5 & 4x^4 \\
2x^5 & 4x^4
\end{array}
\right)
\end{array}.
$$

According to the product, there are three ways to travel from Montréal to Pittsburgh (one route costing four and two routes costing five), there are two ways to travel from Montréal to Washington (both costing five), there are four ways to travel from Boston to Pittsburgh (all costing four) and four ways to travel from Boston to Pittsburgh (all costing four).

However, the number of victory points may be more interesting than the number of cards used. Each route earns victory points based on the number of cards used to claim it, as shown in Table 4.1. To calculate victory points instead of costs, one uses the victory point value in the exponent instead of the number of cards used. This will work because the victory points are added when routes are chained together. The only change here is that a length three route is worth four victory points. The variable has also been changed from x to v (for victory points),

$$
\begin{pmatrix}
v^2 & 2v^2 \\
0 & 2v^2
\end{pmatrix}
\begin{pmatrix}
v^2 & 0 \\
v^4 & 2v^2
\end{pmatrix}
=
\begin{array}{c}
\quad\quad\quad Mon \quad\quad Bos \\
\begin{array}{c} Pitt \\ Wash \end{array}
\left(
\begin{array}{cc}
v^4 + 2v^6 & 4v^4 \\
2v^6 & 4v^4
\end{array}
\right)
\end{array}.
$$

From this, it can be seen that there are three routes from Montréal to Pittsburgh. One awards the player four victory points, and two award the player six.

This chapter's opening question asked about routes from Montréal to Los Angeles can be answered by constructing a weighted adjacency matrix for

Ticket To Ride and raising it increasing powers until the entry associated with a route from Montréal to Los Angeles is non-zero. The fifth power of the matrix has a zero in the Montréal–New York position, but the sixth power results in the polynomial

$$p(v) = v^{55} + v^{53} + 4v^{52} + 5v^{50} + 4v^{47} + v^{44} + 3v^{42} + 2v^{39} + 2v^{37}.$$

From this, there are 23 routes, as this is the sum of the coefficients (which is also the value of $p(1)$). The highest-scoring route is worth 55 points, while the lowest is only worth 37 points. This leaves some questions that can be explored. How much do these routes cost, which cards are required, and what are the routes?

Determining how many cards are required for each route can be calculated by including factors for card requirements (the expressions in x) with the factors for the victory points (the expressions in v). When multiplying expressions with different variables, the variables will not interact. Returning to the small case shown in Figure 4.9, it is important to note that the expression corresponding to the two routes from Boston to New York is

$$\underbrace{2}_{\text{number of routes}} \times \underbrace{x^2}_{\text{cost}} \times \underbrace{v^2}_{\text{victory points}} = 2x^2v^2$$

and not $(2x^2)(2v^2) = 4x^2v^2$ which would indicate four routes. One can compute each route's cost and victory points with the product

$$\begin{pmatrix} x^2v^2 & 2x^2v^2 \\ 0 & 2x^2v^2 \end{pmatrix} \begin{pmatrix} x^2v^2 & 0 \\ x^3v^4 & 2x^2v^2 \end{pmatrix} = \begin{matrix} Pitt \\ Wash \end{matrix} \begin{pmatrix} Mon & Bos \\ x^4v^4 + 2x^5v^6 & 4x^4v^4 \\ 2x^5v^6 & 4x^4v^4 \end{pmatrix}.$$

The same calculation for six-hop routes from Montréal to Los Angeles in the full *Ticket To Ride* matrix results in

$$p(v, x) = v^{55}x^{27} + v^{53}x^{27} + 4v^{52}x^{26} + 5v^{50}x^{26}$$
$$+ 4v^{47}x^{25} + v^{44}x^{23} + 3v^{42}x^{23} + v^{39}x^{23} + v^{39}x^{22} + 2v^{37}x^{22}.$$

Notice that the number of routes is equal to $p(1, 1)$, the polynomial $p(v, 1)$ picks out the number of victory points, and the polynomial $p(1, x)$ picks out the number of cards required.

One feature that can be seen here is that the four routes earning 52 victory points and costing 26 cards are more efficient ways to earn points (2 points per card) than the route earning 53 victory points and costing 27 cards (1.96 points per card).

Until now, all cards have been treated as interchangeable (placing the number of cards in the exponent of the variable x). However, there is no requirement that all cards use the same variable. A different variable can be used for each card color to track the colors of the cards required. Again, referring to Figure 4.9, the variables in the legend are used to encode the colors of the cards required for each route. When this is done for Figure 4.9, the matrix describes the cards required for each route:

$$\begin{pmatrix} x^2 & w^2 + g^2 \\ 0 & b^2 + t^2 \end{pmatrix} \begin{pmatrix} x^2 & 0 \\ b^3 & y^2 + r^2 \end{pmatrix}$$

$$= \begin{array}{c} \\ Pitt \\ Wush \end{array} \begin{pmatrix} \overset{Mon}{x^4 + w^2 b^3 + g^2 b^3} & \overset{Bos}{w^2 y^2 + w^2 r^2 + g^2 y^2 + g^2 r^2} \\ b^5 + b^2 t^2 & b^2 y^2 + b^2 r^2 + t^2 y^2 + t^2 r^2 \end{pmatrix}$$

According to this matrix, there are three ways to travel from Montréal to Pittsburgh: with four wild cards, with two white cards and three blue cards, or with two green cards and three blue cards.

Doing the computation for the full *Ticket To Ride* game, it is possible to determine the various color costs for the Montréal to Los Angeles route, and the cost for the highest scoring route is $g^4 p^6 t^6 w^5 x^6$, indicating that the route requires four green, six pink, six tangerine, five white, and six wild cards. Returning to the original map for *Ticket To Ride*, this color combination can be used to find the route: Montréal–Toronto–Duluth–Helena–Denver–Phoenix–Los Angeles.

To determine the route taken without needing to return to the map, each city on the map can be assigned a variable, and the matrix product will result in a list of cities visited for each route. This idea is computationally expensive and is explored in supplemental material (see Appendix A).

Raising the matrix to the seventh power, it is discovered that there are 356 routes using seven hops, and at most 73 victory points possible with the route Montréal–Toronto–Duluth–Helena–Denver–Oklahoma City–El Paso–Los Angeles. Allowing eight hops, there are 3763 routes, but many are not paths. For example, this count includes routes such as Montréal–Toronto–Montréal–Toronto–Duluth–Helena–Denver–Phoenix–Los Angeles.

For many games, walks which are not paths are not relevant. In *Fury of Dracula* (p.63), hunters gain no advantage by returning to the same space more than once in a single turn. In *Ticket To Ride*, loops in routes do not increase scores. Because the adjacency matrix counts the number of walks, it will overcount the possibilities. If city names are encoded in the matrix, one can remove the cycles by removing all entries in the resulting expression that contain locations to powers higher than one. However, the number of computations quickly becomes overwhelming, even for a computer.

This section has primarily focused on games, where pieces are placed on the vertices or edges of a graph. In *War of the Ring*, based on the J.R.R. Tolkien books, one player takes the role of the free people attempting to destroy the One Ring, and the other takes the role of the shadow forces attempting to conquer Middle Earth. In this game, each region may contain many different pieces, so placement on vertices would be challenging to manage in gameplay.

However, converting the region graph to its dual (where a vertex represents each region) can help answer questions about routes between locations (such as between Rivendell and Mordor).

4.3 DIRECTED GRAPHS

While matrices work well in determining possible walks, they have difficulty distinguishing which of these walks are paths. Finding all paths in a graph is computationally very difficult, but there are methods to find all paths that start from a given vertex. A frequently used method is a depth-first search, where a path is extended for as long as possible. When the path can no longer be extended because no edges can be followed to continue the path, the method will backtrack to the most recent vertex with edges that can be used to extend a path.

Consider the graph in Figure 4.9 with the starting vertex of Montréal. A depth-first search results in extending paths, as shown in Figure 4.11 (reading from top to bottom). Starting with Montréal, the path can be extended.

$$\text{Montréal—Toronto—Pittsburgh}\overset{ww}{\text{—}}\text{New York}\overset{rr}{\text{—}}\text{Boston.}$$

The path cannot continue at this point as the only remaining edge returns to New York. Backtracking to New York and then continuing yields the three paths:

$$\text{Montréal—Toronto—Pittsburgh}\overset{ww}{\text{—}}\text{New York}\overset{yy}{\text{—}}\text{Boston}$$

$$\text{Montréal—Toronto—Pittsburgh}\overset{ww}{\text{—}}\text{New York}\overset{tt}{\text{—}}\text{Washington}$$

$$\text{Montréal—Toronto—Pittsburgh}\overset{ww}{\text{—}}\text{New York}\overset{bb}{\text{—}}\text{Washington.}$$

At this point, no other edges are leaving New York that would extend the path. So, the algorithm backtracks to Pittsburgh and finds the following four paths:

$$\text{Montréal—Toronto—Pittsburgh}\overset{gg}{\text{—}}\text{New York}\overset{rr}{\text{—}}\text{Boston}$$

$$\text{Montréal—Toronto—Pittsburgh}\overset{gg}{\text{—}}\text{New York}\overset{yy}{\text{—}}\text{Boston}$$

$$\text{Montréal—Toronto—Pittsburgh}\overset{gg}{\text{—}}\text{New York}\overset{tt}{\text{—}}\text{Washington}$$

$$\text{Montréal—Toronto—Pittsburgh}\overset{gg}{\text{—}}\text{New York}\overset{bb}{\text{—}}\text{Washington.}$$

This process repeats until no unfollowed edges leave Montréal. This process can be visualized as building the graph shown in Figure 4.11. Here, the arrowheads added to the edges indicate that these edges can only be traversed in one direction.

Notice that each vertex in Figure 4.11 represents a path from Montréal to the vertex with the same label in Figure 4.9. There are three ways to arrive at Toronto (one one-hop route and two three-hop routes) because there are three vertices in the new graph labeled Toronto (marked with a double border).

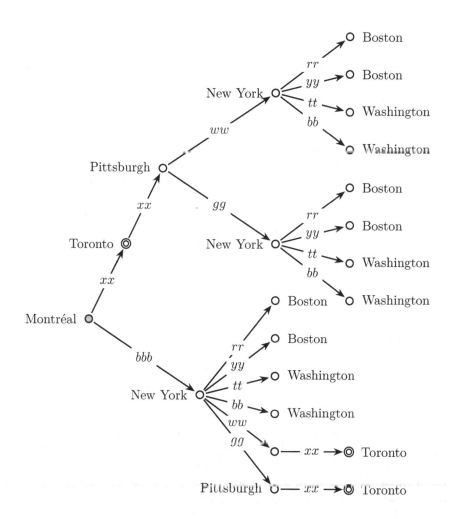

Figure 4.11: Paths from Montréal in the Simplified Map for *Ticket To Ride*.

A depth-first search for paths from a starting vertex requires selecting the starting vertex. As a result, it is only effective at finding paths originating at that particular vertex. In contrast, the adjacency matrix method from Section 4.1 simultaneously finds walks from all vertices. Implementing a depth-first search for paths from Montréal to Los Angeles in a computer algebra system finds a total of 2751 eight-hop paths from Montréal to Los Angeles. Finding all paths from Montréal to Los Angeles (of any length) is much more difficult. In the worst case scenario, a graph with n vertices will have $\lfloor (n-2)!e \rfloor$ paths, where $\lfloor x \rfloor$ is the floor of x (the largest integer less than

x) and e is Euler's constant[2] (approximately 2.7). Because this number grows at a factorial rate, it is unreasonable to be able to pursue this technique for graphs with a large number of vertices, like the *Ticket To Ride* graph.

Connected graphs containing no cycles (paths from a vertex back to itself) are trees. In a tree, every two vertices are connected by a unique path. While the graph in Figure 4.9 is not a tree, the graph in Figure 4.11 is a tree. A rooted tree is a tree, where one of its vertices has been selected as the root. The graph in Figure 4.11 is a rooted tree with its root at Montréal. This book will use the convention that the root is filled with gray, as shown in Figure 4.11, and the direction of the arrows can be inferred from the location of the root. In later graphs, the arrowheads will be omitted from the tree in these cases.

More generally, a graph where edges have a direction is a directed graph. This is indicated by placing arrowheads on the edges to indicate the direction. When identifying walks, paths, and cycles in a directed graph, the movement from vertex to vertex must match the direction of the arrow.

Partial Orders

Rooted trees also effectively describe relations such as the less-than relation. This can make it easier to visualize these relations or to see other patterns that may be difficult to detect without the graphical representation.

In many games, players must choose between options that are not directly comparable. For example, in *Project L*, players use polyominoes to fill shapes. Once a shape is filled, the player receives a reward depending on the shape filled. The reward consists of a new polyomino (so the player can fill more shapes), victory points, or both. Sometimes, the player can quickly determine the relative worth of filling in one shape versus another. However, at other times, it is unclear if one reward is more valuable. In Figure 4.12, the player has a square tetromino to place and can choose to complete either the left shape (earning two points and a new square tetromino) or the right shape (earning five points and a new monomino). Early in the game, gaining the new tetromino may be more valuable (and the five points can be earned later), but later in the game, the five points may be more valuable.

The rewards in *Project L* are related by a strict partial order. The properties of strict partial orders are listed in Strict Partial Order Properties. A familiar example of a strict partial order is the relation of one person being an ancestor of another person.

---

[2]Leonhard Euler was an exceptionally prolific eighteenth-century Swiss mathematician who did groundbreaking research in several fields of mathematics with 866 publications over his lifetime.

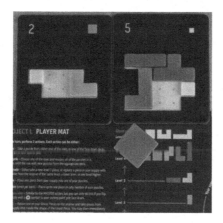

Figure 4.12: Puzzles in *Project L*.[3]

Strict Partial Order Properties: A relation, "precedes," satisfying these three properties is a strict partial order.

Irreflexivity: No object precedes itself.

Asymmetry: If one object precedes a second object, the second object does not precede the first.

Transitivity: If one object precedes a second object, and the second object precedes a third object, then the first object precedes the third object.

Given a strict partial order, the associated partial order is the relation of either preceding or being equal. Sets that have a partial order (strict or not) are referred to as posets (short for "partially ordered set"). If a precedes c and there is no element b that is between a and c, then c covers a. In the ancestor relation, a parent covers their children. A Hasse diagram[4] is a graph used to display a partial order visually and has a vertex for every object, and arrows are drawn from each element to every element that covers it. In the ancestry example, a Hasse diagram would be a family tree.

In the case of *Project L*, there will be a vertex for every possible reward. The vertices will be labeled as ordered pairs with the number of victory points first and the number of cells of the earned tile second. More victory points are better than fewer victory points, so $(0, 2)$ covers $(1, 2)$, which covers $(2, 2)$. Similarly, larger tiles are better than smaller tiles, so $(1, 2)$ covers $(1, 3)$, which

---

[4]A Hasse diagram is named after Helmut Hasse, a twentieth-century German mathematician who used these diagrams frequently as part of his research.

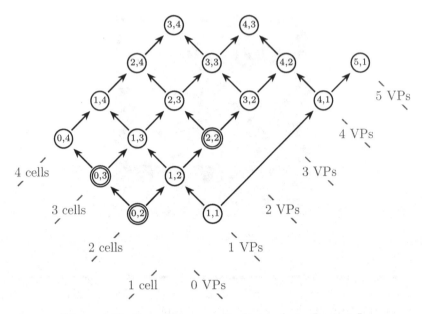

Figure 4.13: Hasse Diagram for Rewards in *Project L*.

covers $(1, 4)$. When constructing the Hasse diagram for the rewards in *Project L*, there is an arrow from $(0, 2)$ to $(1, 2)$ and from $(1, 2)$ to $(1, 3)$. There is no arrow from $(0, 2)$ to $(1, 3)$ because the reward of $(1, 2)$ is between them. One reward is less valuable if there is a path (following the direction of the arrows) from the first reward's vertex to the second reward's vertex. So $(0, 2)$ precedes $(2, 2)$, but neither of $(0, 3)$ nor $(2, 2)$ precedes the other (these rewards are marked with a double border for easy identification in Figure 4.13).

As the game progresses, players usually shift from collecting more pieces to collecting more victory points, moving from the left of the graph to the right of the graph.

However, there is another way to organize the tiles in the game: the size of the shape to be filled on the tile. Since larger shapes are more challenging to fill, they have been placed higher in the graph in Figure 4.14. In some cases, the same reward is available on tiles of different-sized shapes, but in these cases, the two sizes differ only by one cell and are placed at the average of their sizes (e.g., the $(3, 4)$ reward is found on tiles of size ten and eleven and is placed at height 10.5). The graph shown in Figure 4.14 results from this arrangement. No arrows point downward and the arrows point upward in all but one case, indicating that the value of the reward is consistent with the shape's size.

In *SCOUT* (p.10), players discard runs and sets of cards, either of which will be called a combo. The player to discard all of the cards in their hand first wins the game. A run of cards is any number of cards (one or more) in

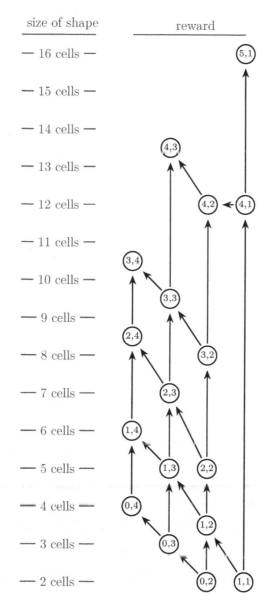

size of shape

reward

— 16 cells —

— 15 cells —

— 14 cells —

— 13 cells —

— 12 cells —

— 11 cells —

— 10 cells —

— 9 cells —

— 8 cells —

— 7 cells —

— 6 cells —

— 5 cells —

— 4 cells —

— 3 cells —

— 2 cells —

Figure 4.14: Rewards by Puzzle Size in *Project L*.

sequence (in either direction). So for example, $(1, 2, 3)$ is a run, as is $(3, 2, 1)$, but $(1, 3, 2)$ is not a run. A set of cards is any collection of cards, all with the same number. Runs and sets are shown in Figure 4.15. On a player's turn, they may play a combo of cards stronger than the combo on the table. The

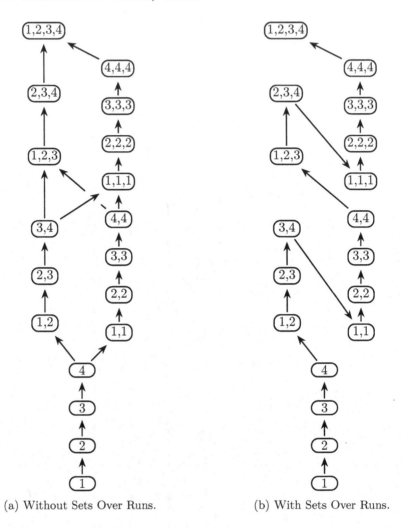

(a) Without Sets Over Runs.   (b) With Sets Over Runs.

Figure 4.15: Strength of Combos in *SCOUT*.

three rules in Combo Strength in *SCOUT* are used to determine the strength of a combo.

Combo Strength in *SCOUT*

More Cards A combo with more cards is stronger than a combo with fewer cards.

Higher Cards If the number of cards and the combo type are the same, a combo with the higher card is stronger than the other.

Sets Over Runs If the number of cards is the same and the combo types differ, a set is stronger than a run.

The first two rules use an already established order (the number of cards and the number on the cards). The third rule makes an arbitrary ranking between sets and runs, and it is reasonable to ask how the game changes if the third rule is not included. Consider a smaller version of $SCOUT$ where the cards have values from one to four (instead of one to ten in the full version), with each value from one to four appearing three times. The Hasse diagram of the strength of a combo without including the **Sets Over Runs** rule is shown in Figure 4.15a. This can be compared to the Hasse diagram with the extra rule shown in Figure 4.15b. Adding the rule has increased the number of sequential combos that can be made above a given combo. Without the rule, there are only seven combos above a $(1, 2)$, whereas, with the rule, there are thirteen combos above a $(1, 2)$.

A strict total order is another relation. It extends the properties of a partial order by requiring that each element be comparable, as described in Strict Total Order Properties on page 83. Again, the term "strict" indicates that the relation is not reflexive.

Strict Total Order Properties: The relation, "precedes," satisfying these three properties is a strict total order.

Irreflexivity: No object precedes itself.

Asymmetric: If an object precedes a second object, then the second object does not precede the first object.

Transitivity: If one object precedes a second object and the second object precedes a third object, then the first object precedes the third object.

Connectivity: If two elements are not equal, then one precedes the other.

The result of adding the final rule in $SCOUT$ is to guarantee that any two combos can be compared, converting the strict partial order (in Figure 4.15a) to a strict total order (in Figure 4.15b). A linear extension of a strict partial order is a strict total order in which if one object is less than a second in the strict partial order, the first object will also be less than the second in the strict total order. This is related to topological sorts in computer science.

Campaigns

Some tabletop games provide a campaign where players play through individual scenarios. For example, in *Pandemic*, players act as emergency workers who work cooperatively to prevent disease outbreaks worldwide. The game *Pandemic Legacy: Season 1* expands this gameplay by connecting twelve scenarios to chronicle a year-long story, with each scenario portraying a month.

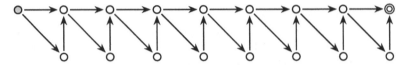

Figure 4.16: Simplified Campaign Map for *Pandemic Legacy: Season 1.*

Pandemic Legacy: Season 1 is what is termed a legacy game, where each scenario extends or changes the rules of the basic game. Players who do well in a scenario will move from one month to the next. However, players can attempt a month again if they do poorly.

Because the sequence of scenarios in campaign games tells a larger story, the order in which the scenarios occur is chosen to develop a coherent story. In these cases, a directed graph can represent the progression through the story. Using these diagrams can help designers ensure the consistency of the story and help players determine the options available to them.

The campaign diagram in Figure 4.16 is similar to that found in *Pandemic Legacy: Season 1*. The campaign begins with the gray vertex and ends in an epilogue (the vertex with a double border). After completing a scenario, players may either continue to the next or repeat the current scenario and then move on, as shown in Figure 4.16. Even though players can have different experiences as they work through the campaign, the structure of the scenarios is relatively constrained. Games with this type of structure are often referred to as being "on rails" as players have little choice over the order in which the scenarios will occur. One advantage of a linear campaign is that the designers have more freedom to craft the overarching story and match the difficulty of the scenarios with the players' experience. While an adjacency matrix is not necessary to determine that there are $2^8 = 256$ paths through the scenarios (as there are eight branching vertices, each with two options), an adjacency matrix confirms this result. It should be noted that these diagrams are not Hasse diagrams, as they contain edges that arise from non-covering relations.

The game *Gloomhaven* is a large-scale campaign set in a high-fantasy world. Players take on the role of adventurers, each adventurer type having a unique deck of ability cards that allow them to defeat enemies and traps as they level up and eventually retire. One innovative feature of this game is that instead of rolling dice to determine whether characters succeed in combat, players draw from a personal modifier deck that can be upgraded during the game. Throughout the game, players are invited to explore an expansive world (with over 100 scenarios) and at any given time may have a multitude of available scenarios available to them. A campaign diagram similar to that for *Gloomhaven* is shown in Figure 4.17. The initial scenario is shaded gray, and the concluding scenarios have double borders. Games with this type of structure are often called "sandbox games" as they allow players to interact with the story relatively free of restrictions. Using an adjacency matrix, one can discover that there are six paths leading to the concluding scenario in the

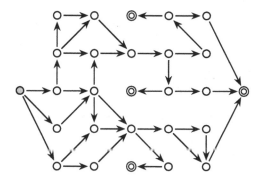

Figure 4.17: Simplified Campaign Map for *Gloomhaven*.

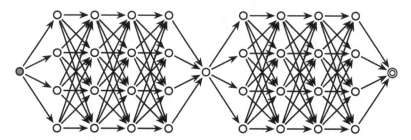

Figure 4.18: Simplified Campaign Map for *Hoplomachus: Remastered*.

bottom row, there are five paths leading to the two other concluding scenarios in the middle column, and there are 22 paths leading to the concluding scenario in the last column.

In between these two types of campaigns are those which establish fixed bottlenecks. For example, in *Hoplomachus: Remastered*, players take on the roles of adventurers in a mythical Greek setting doing battle with four large enemies. The characters travel the world and engage in minor skirmishes between these more significant battles. As a result of this structure, the campaign diagram has similar features to the one shown in Figure 4.18 (which has one bottleneck). The Power Rule can compute the number of paths to the concluding scenario as $4^8 = 65\,536$ as there are eight branching vertices, each with four options, which is confirmed by a calculation using an adjacency matrix.

CHAPTER 5

Probability

Figure 5.1: Components from *Oathsworn: Into the Deep Wood.*[1]

In the game *Oathsworn: Into the Deep Wood*, players act the part of adventurers battling against enemies in the Deepwood in a grimdark setting. In combat, players may roll dice or draw cards, shown in Figure 5.1, when determining how much damage they do to their enemies. The decks in the game are designed so that the probability of drawing a number from a fresh deck of cards matches the probability of rolling the same number on a die.

How significantly does the choice of drawing over rolling affect the total damage done?

---

[1]Oathsworn is a trademark of Shadowborne Games LLC, All Rights Reserved 2023.

 DOI: 10.1201/9781003383529-5

INTRODUCTION

Probability plays a significant role in many games, and, as a result, this is the longest chapter of the book. Here, the focus is on discrete distributions (with a finite number of outcomes). One commonly used distinction between Eurogames and American-style games is the incorporation of randomness. In Eurogames, a random process often occurs before a player makes any decision, such as shuffling a deck of cards before dealing them out (this is known as input randomness). In American-style games, often a player will decide on an action and then a random process will determine the effect of that decision, such as when a player rolls a die to determine if they succeed at a task (this is known as output randomness).

The primary forms of randomization in tabletop games are shuffling cards and rolling dice. This chapter covers the distributions associated with these activities. It also covers the practical difference between rolling dice and drawing cards in the context of the game *Oathsworn: Into the Deep Wood*, where players may choose to use either method in combat.

The section closes with a discussion of Markov processes. These methods effectively determine the outcome of a process in which a probability distribution determines every transition. Such methods effectively model many combat simulations and the roll-and-move mechanism found in some games.

5.1 BASIC PROBABILITY

In the *Zombicide* series of games, players adopt the role of characters fighting zombies. The *Zombicide: Black Plague* game in the series is set in a medieval fantasy setting. If a character is equipped with a short bow, the player will roll a six sided die and hit on a roll of three or more. What is the probability of scoring a hit when attacking with a short bow?

Much of the probability calculations in tabletop gaming come from applying the counting rules from Chapter 1. The first step is determining an exhaustive, mutually exclusive collection of possible outcomes. Exhaustive means that every possible outcome is included in the list, and mutually exclusive means that two outcomes in the collection cannot both occur. An exhaustive and mutually exclusive collection is a sample space. An experiment occurs when one outcome from a sample space is randomly selected. A sample space where all results are equally likely is a uniform distribution. If the sample space consists of n equally likely outcomes, then the probability that any one of them occurs in an experiment is $1/n$. For die rolls, the die is fair if each roll is equally likely. For a fair six-sided die, the probability of rolling a two would be $1/6$ as it is one of six equally likely outcomes. This is written as $P[\text{roll a 2}] = 1/6$. Throughout this book, the focus will be on sample spaces of equally likely outcomes unless otherwise stated. Returning to *Zombicide: Black Plague*, when attacking with a short bow, a good sample space for this would be the equally likely outcomes

$$\{\text{roll a } 1, \text{roll a } 2, \text{roll a } 3, \text{roll a } 4, \text{roll a } 5, \text{roll a } 6\}.$$

On the other hand, the sample space $\{\text{hit}, \text{miss}\}$ would be a poor choice since the probability of hitting is not the same as the probability of missing. An event consists of a subset of the sample space. An event from the sample space above might be "score a hit" which is the set $\{\text{roll a } 3, \text{roll a } 4, \text{roll a } 5, \text{roll a } 6\}$. The probability of an event occurring can be calculated by summing the probability of each outcome. The probability of scoring a hit with the short bow is approximately 67%:

$$P[\text{roll a hit}] = P[\text{roll a } 3, 4, 5, \text{ or } 6]$$
$$= P[\text{roll a } 3] + P[\text{roll a } 4] + P[\text{roll a } 5] + P[\text{roll a } 6]$$
$$= \frac{1}{6} + \frac{1}{6} + \frac{1}{6} + \frac{1}{6} = \frac{2}{3} \approx 0.67. \quad (\dagger)$$

An alternative description of this calculation for sample spaces of equally likely outcomes is to divide the number of outcomes in the event by the number of outcomes in the sample space. Here this would give $P[\text{roll a hit}] = {}^4/_6 = {}^2/_3$.

Intersection of Events

Many complex events can be decomposed into simpler events. For example, the result of rolling two six-sided dice can be decomposed into two events of rolling a six-sided die. Each roll has the same sample space, but rolling the pair results in the sample space S,

$$S = \{ \quad (1,1), \quad (1,2), \quad (1,3), \quad (1,4), \quad (1,5), \quad (1,6),$$
$$(2,1), \quad (2,2), \quad (2,3), \quad \cdots$$
$$\cdots, \quad (6,4) \quad (6,5), \quad (6,6) \quad \}.$$

While this is correct, it is easier to describe the set as $S = \{1, 2, 3, 4, 5, 6\} \times \{1, 2, 3, 4, 5, 6\}$ and the Product Rule calculates the total number of outcomes: $6 \times 6 = 36$. Returning to *Zombicide: Black Plague*, a character may be equipped with a sword. In this case, the player can roll a six-sided die twice (or two six-sided dice once) and score a hit for each roll of four or more. How likely is the character to score two hits?

In this case, the two experiments are the two rolls of the die, and the goal is to determine $P[\text{first roll is a hit and second roll is a hit}]$. An event of this type is the intersection of the component events. The probabilities of these events can be calculated by considering the probabilities of each component event separately and then combining them using the Product Rule. However, the calculation must carefully account for hidden connections between the events.

Two events, A and B, are independent if

$$P[A \text{ and } B] = P[A] \times P[B].$$

In the case of rolling two dice, the standing assumption is that the rolls are independent, and this probability is 25%:

$$P[\text{two hits}] = P[\text{first hit}] \times P[\text{second hit}] = \frac{1}{2} \times \frac{1}{2} = \frac{1}{4}.$$

If the events are not independent, the calculation is more complicated. Given two events, A and B, the conditional probability of A given B, written $P[A \mid B]$, is the probability that A occurs when restricted to outcomes where B occurs. Using this notation, the probability of compound events can be determined with the Product Rule for Probability.

Product Rule for Probability:

$$P[A \text{ and } B] = P[B] \times P[A \mid B]$$
$$= P[A] \times P[B \mid A]$$

If A and B are independent, then $P[A \mid B] = P[A]$ and $P[B \mid A] = P[B]$. One way to understand this is that if two events are *not* independent, then learning that one has occurred allows a person to adjust the chance that the other will occur in their calculations.

Rolling dice often results in independent events. For example, rolling a six-sided die twice results in two independent events: if one learns that the first roll is a six, they cannot adjust the probability that the second dice would also be a six:

$$\frac{1}{6} = P[\text{second dice rolls a six}]$$
$$\frac{1}{6} = P[\text{second dice rolls a six} \mid \text{first dice rolls a six}].$$

On the other hand, drawing cards often results in non-independent events. For example, drawing two cards from a standard deck of cards results in two non-independent events: if one learns that the first card is an Ace, they can adjust the probability that the second card will also be an Ace:

$$\frac{4}{52} = P[\text{second card drawn is an Ace}]$$
$$\frac{3}{51} = P[\text{second card drawn is an Ace} \mid \text{first card drawn is an Ace}].$$

An event is impossible if it has probability zero. For example, rolling a seven would be impossible if one is rolling a standard six-sided die. One extreme case of dependence is when two events are mutually exclusive, meaning that their intersection is impossible. If A and B are mutually exclusive, $P[A \mid B] = 0$.

Union of Events

More complicated is the situation where either one or another event occurs. This is referred to as the union of the two events. In the most straightforward situation, the two events are mutually exclusive. If this is the case, one can add the probabilities of the two components together. However, it is often more complex.

The probability that a character in *Zombicide: Black Plague* scores at least one hit while attacking with a sword is the probability that the first or the second attack results in a hit. Adding these two probabilities leads to a probability of 100%:

$$P[\text{first hit}] + P[\text{second hit}] = \frac{1}{2} + \frac{1}{2} = 1.$$

If this were correct, a character with a sword would never miss. However, from experience, characters with swords miss. What has happened is that the possibility that both die rolls hit was counted *twice*, first in $P[\text{first hit}]$ and then again in $P[\text{second hit}]$. To accommodate this, the extra counting must be subtracted as in the Inclusion-Exclusion Principle to arrive at a probability of 75%:

$$P[\text{at least one hit}]$$
$$= P[\text{first hit}] + P[\text{second hit}] - P[\text{both hit}]$$
$$= \frac{1}{2} + \frac{1}{2} - \frac{1}{4} = \frac{3}{4} = 0.75.$$

The Sum Rule for Probability handles the most common case of two events. The union of more events will follow the Inclusion-Exclusion Principle pattern with an alternating sum.

Sum Rule for Probability:

$$P[A \text{ or } B] = P[A] + P[B] - P[A \text{ and } B].$$

Probability Distribution Trees

Ensuring every outcome is counted precisely once when tracking compound events can be difficult. One way to track the outcomes is to use a tree, where each edge corresponds to a component event. For example, if a character is attacking with a repeating crossbow in *Zombicide: Black Plague*, the player rolls three dice and scores a hit for each roll of five or greater. What is the probability of achieving exactly two hits?

While this can be computed using the Inclusion-Exclusion Principle, the calculation can also be done using a rooted tree, referred to as a probability

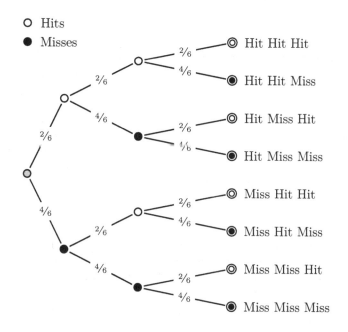

Figure 5.2: Attacking in *Zombicide: Black Plague*.

tree. The root vertex is the state before any die has been rolled. When the first die is rolled, there are two possible states: it rolled a "Hit" (with probability $2/6$) or a "Miss" (with probability $4/6$). This provides two edges from the root to these two states. From each of these states, there are two possibilities for the second die roll, resulting in four branches. From each of these four results, there are two possibilities for the third die roll, resulting in eight final branches. When doing this, the probabilities are included as labels on the edges, as shown in Figure 5.2.

From this, the probability of any outcomes at the leaves can be obtained using Product Rule for Probability. For example, the probability of only hitting on the first and third die (Hit Miss Hit) is given as $(2/6)(4/6)(2/6) = 16/216$. The calculations in the tree can be checked by summing all branches to obtain one. To determine the probability of exactly two hits, the probabilities from the branches "Hit Hit Miss," "Hit Miss Hit," and "Miss Hit Hit" are summed to arrive at approximately 22%:

$$\left(\frac{2}{6}\right)\left(\frac{2}{6}\right)\left(\frac{4}{6}\right) + \left(\frac{2}{6}\right)\left(\frac{4}{6}\right)\left(\frac{2}{6}\right) + \left(\frac{4}{6}\right)\left(\frac{2}{6}\right)\left(\frac{2}{6}\right) = \frac{48}{216} \approx 0.22.$$

A more sophisticated method to handle this example will be developed in Section 5.3.

5.2 PARAMETERS

Frequently, the raw outcome of an experiment is only important as a means to derive a numerical value. For example, in *Zombicide: Black Plague* (p.87), the exact values on the dice are only a means to determine the number of hits. A function from the sample space to the numerical values of interest is a random variable. So, in the *Zombicide: Black Plague*, the random variable would map the dice rolls to the number of hits. Random variables are often represented by a capital letter. For example, the variable H may represent the number of hits scored with a short bow in *Zombicide: Black Plague*, and the fact that the probability of achieving one hit is two-thirds (computed in Equation (†) on page 88) would be written as

$$P[H = 1] = \frac{2}{3}.$$

A parameter is a number that describes some feature of a random variable, and the primary parameters considered here provide information about the "average" value of the random variable. Not every parameter is appropriate for every collection of data. Before calculating a particular parameter for a random variable, one should determine if the parameter's value is meaningful. Determining which parameters apply to which random variables is often done by considering the type of data being analyzed.

Data Types

Nominal data refers to data that cannot be ranked. A variable that encodes a player's color preference would be an example of nominal data. It may be convenient to encode this data as an integer. For example, "prefers red" may be encoded as a one, "prefers green" may be encoded as a two, and so forth. These values cannot be compared or averaged, as there is no meaningful interpretation to claim that the average player prefers the color associated with the encoding 1.5.

Ordinal Data only indicate an objects' rankings but not the objects' relative values. For example, *War* may be a player's favorite game, followed by *Candy Land*, and then *Monopoly*, and then *Yahtzee*. Even if these preferences are encoded as one through four respectively, the data do not support the conclusion that the gap between the player's preference for *Candy Land* over *Monopoly* is the same as the gap between their preference for *War* over *Candy Land*.

Interval Data are data for which subtraction makes sense in the game context. In a game of *Twilight Struggle*, players take the roles of the US and the USSR during the Cold War and take turns playing cards that represent historical actions to move a token in one-step intervals between a US victory (which could be assigned the value +20) and a USSR victory (which could be assigned the value −20). Shifting the token one step takes the same effort no matter where the token starts (so, for example, moving from +3 to +2 is the same effort as moving from -5 to -6). In this case, subtraction makes sense: it indicates the steps required to move the token between the two scores. However, division is more problematic. A score of $+2$ is not "twice as good" for the US player as having a score of $+1$.

Ratio Data are data for which division makes sense in the game context. In a game like *Zombicide: Black Plague*, landing two hits is twice as good as landing one hit, so dividing the values here has meaning. As a result, the number of hits in this game is an example of ratio data.

Mean and Median

When working with ordinal data, it makes more sense to describe the center of the data using the median. The median is chosen so that half of the data are above the median while half are below the median. If there are an odd number of values, the median is the middle value (by rank):

$$\underbrace{x_1 \leq x_2 \leq \cdots \leq x_n}_{n \text{ values}} \leq \underbrace{x_{n+1}}_{\text{median}} \leq \underbrace{x_{n+2} \leq \cdots \leq x_{2n+1}}_{n \text{ values}}.$$

If there are an even number of values, the median is between the two middle values. If it makes sense to compute the midpoint of two data points, the median is the midpoint of the middle two data points:

$$\underbrace{x_1 \leq x_2 \leq \cdots \leq x_n}_{n \text{ values}} \leq \underbrace{\frac{x_n + x_{n+1}}{2}}_{\text{median}} \leq \underbrace{x_{n+1} \leq \cdots \leq x_{2n}}_{n \text{ values}}.$$

It may not make sense to compute the midpoint of two values of ordinal data. In the example of the gamer listing their game preferences, there is no midpoint between the games *Candy Land* and *Monopoly*. Nonetheless, saying the median is between them is still helpful in determining whether a game is in the upper or lower half of the collection. One way to think of the median is that over repeated experiments, roughly half of the time, the results will be above the median, and roughly half of the time, the results will be below the median. Returning to the ranking example and assuming that all games in the collection are equally likely to be selected, the median says that roughly half the time, the player will play a game they prefer more than *Monopoly*.

When working with interval or ratio data, the expected value (also referred to as the mean) can be used. Given a random variable X, the expected value of X is defined using the formula below, where x ranges over all possible values of X.

$$E[X] = \sum x P[X = x].$$

The notation μ will be used for the expected value in places, where the random variable is clear from context. When μ is written as a fraction p/q, this can be interpreted as saying that for every q experiments, the sum of the outcomes will be p, on average.

When rolling dice in *Zombicide: Black Plague*, the player is not interested in the sum or average of the rolls but in tracking the number of successes. If H is the number of hits using a repeating crossbow (see Figure 5.2), then $E[H] = 1$:

$$E[H] = 3 \times \frac{8}{216} + 2 \times \frac{48}{216} + 1 \times \frac{96}{216} + 0 \times \frac{64}{216} = 1.$$

Interpreting this as a fraction, $1/1$, indicates an expectation of one hit for every attack on average. Similar hit mechanisms are found in many games, including *Hoplomachus: Remastered* (p.85) and *War of the Ring* (p.75).

One very nice property of the expected value is that it is linear. In particular, the expected value of repeating an experiment n times can be determined from the expected value of performing the experiment once and multiplying by n.

Linearity of Expectation: If X and Y are random variables and n is a positive integer, then

$$E[X + Y] = E[X] + E[Y]$$
$$E[nX] = nE[X].$$

Exploding Dice

One mechanism found in games like *Zombicide: Black Plague* and *Oathsworn: Into the Deep Wood* (p.86) is called exploding dice. This is when rolling a result on one die allows a player to roll a bonus die. Characters in *Zombicide: Black Plague* can use the disintegrate spell to attack zombies, which allows the player to roll three six-sided dice and score a hit on a roll of either a five or a six. Furthermore, the player may roll a bonus six-sided die to score more hits for every six rolled (either as the original or as a bonus roll). What is the expected number of hits in one attack with this weapon?

Attempting to compute the expected value of rolling one die using the definition of the expected value leads to a non-terminating sum since any non-negative number of hits is possible. This is an infinite series, and calculus is necessary to determine whether it results in a well-defined number (it does). When an infinite series results in a well-defined number, finding the value is sometimes possible even without knowing how to calculate the value of an infinite series. Using the variable μ to represent the expected value of a single die roll, it is possible to write μ in terms of itself. Rolling a single die has a $4/6$ probability of resulting in zero hits (rolling a one through four), a $1/6$ probability of resulting in one hit (rolling a five), and a $1/6$ probability of resulting in one hit and another μ hits from the bonus roll. Solving the equation

$$\mu = \left(\frac{4}{6}\right)(0) + \left(\frac{1}{6}\right)(1) + \left(\frac{1}{6}\right)(1 + \mu).$$

for μ arrives at $\mu = 2/5$. When rolling three dice, the expected number of hits is $6/5$ per attack, or six hits for every five attacks, on average.

Notice that this calculation relies on the fact that μ is a well-defined value. For example, the following calculation demonstrates what can happen if this assumption is unmet. Assume that μ is the sum of all powers of two:

$$\mu = 1 + 2 + 4 + 8 + 16 + \dots.$$

Using this equation, μ can be written in terms of itself:

$$\mu = 1 + 2(1 + 2 + 4 + 8 + \dots) = 1 + 2\mu.$$

Solving for μ in this equation leads to $\mu = -1$, a clearly absurd result.

5.3   BINOMIAL DISTRIBUTION

Returning to *Zombicide: Black Plague* (p.87), a character could be equipped with a great sword that allows one to attack with five six-sided dice and score a hit for each roll of five or greater (allowing for between zero and five hits with a single attack). While a tree can be constructed as in Figure 5.2 to compute the mean number of hits, this would be tedious. Keeping the tree in mind, a more efficient way to determine the answer can be developed. From each vertex of the tree, a single die is rolled with a $2/6$ probability of a Hit (moving along the upward branch) and a $4/6$ probability of a Miss (moving along the downward branch). At the end of the process, there will be one Hit for every upward movement and one Miss for every downward movement.

Achieving five hits requires upward movement at every vertex, so only one leaf corresponds to this event. So $P[5 \text{ hits}] \approx 0.4\%$:

$$P[5 \text{ hits}] = \left(\frac{2}{6}\right)^5 \approx 0.004.$$

To achieve exactly four hits requires upward movement at four vertices and downward movement at one vertex. There are $C(5, 4)$ leaves corresponding to this (the selection of which four of the five vertices were upward movement). Each of those leaves will occur with probability $(2/6)^4 (4/6)^1$ as it is the result of four upward movements (with probability $2/6$ each) and one downward movement (with probability $4/6$). So, $P[4 \text{ hits}] \approx 4.1\%$:

$$P[4 \text{ hits}] = \binom{5}{4}\left(\frac{2}{6}\right)^4\left(\frac{4}{6}\right)^1 \approx 0.041.$$

Similar reasoning leads to the general formula

$$P[k \text{ hits}] = \binom{5}{k}\left(\frac{2}{6}\right)^k\left(\frac{4}{6}\right)^{5-k}.$$

A formula that determines the probability of each value of a random variable is known as a probability mass function (abbreviated to PMF). The formula above is the PMF for the distribution of hits with the great sword.

This distribution is an example of a binomial distribution. In a binomial distribution, an experiment is repeated a fixed number of times, each referred to as a trial. The trials are assumed to be independent (so previous outcomes do not affect the probabilities associated with the current outcome). A specific event, referred to as a success, is tracked to determine how frequently it happens during the trials. Trials that do not result in success are said to result in failure. In the *Zombicide: Black Plague* example, each die roll is a trial, and each hit is a success. Reasoning similar to the example results in the general formula in the Binomial Distribution Properties on page 97.

The mean of a binomial distribution can be calculated using the PMF of the binomial distribution along with the Binomial Theorem. Here, the fact that $C(n, k) = 0$ if $k < 0$ or $k > n$ allows sums over all values of k, including those outside of the range $0 \le k \le n$. If X has a binomial distribution with n trials and a probability of success of p, $E[X] = np$.

$$E[X] = \sum_k k \binom{n}{k} p^k (1 - p)^{n-k}$$

$$= \sum_k k \frac{n!}{k!(n-k)!} p^k (1 - p)^{n-k}$$

$$= \sum_k np \frac{(n-1)!}{(k-1)!(n-k)!} p^{k-1} (1 - p)^{n-k}$$

$$= np \sum_k \frac{(n-1)!}{(k-1)!(n-k)!} p^{k-1} (1 - p)^{n-k}.$$

Introduce the following indices: $m = n - 1$, $j = k - 1$. As k runs over all integers, so does j and notice that $m - j = (n - 1) - (k - 1) = n - k$ so the last expression can be rewritten as

$$E[X] = np \sum_j \frac{m!}{j!(m-j)!} p^j (1 - p)^{m-j}.$$

With the variable change, it is clear that the summation matches the one found in the Binomial Theorem,

$$E[X] = np(p + (1 - p))^m = np(1)^m = np.$$

The mean value here may match intuition. For example, in the case of *Zombicide: Black Plague*, if six six-sided dice are rolled, each of which hit with probability 2/6, the expectation of 6 (2/6) = 2 hits is reasonable. It is nice that intuition is backed up by mathematics. As a practical use in the game, a character may be given the option to choose between the sword, the hand crossbow, and the great sword. The sword (one die, hits on three or greater) will average two hits every three attacks. The hand crossbow (two dice, hits on three or greater) will average four hits every three attacks. The great sword (five dice, hits on a five or greater) will average five hits every three attacks.

The results related to the binomial distribution are summarized in the Binomial Distribution Properties.

Binomial Distribution Properties: The probability of k successes for an n-trial binomial distribution with the probability of success equal to p is

$$P[k \text{ successes}] = \binom{n}{k} p^k (1 - p)^{n-k}.$$

The expected value of this distribution is $\mu = np$.

The domains and supports of the probability mass functions, P, would be included in a more complete discussion of probability. The domain of a function is all values for which the function provides an output. The integers are the domain of the probability mass functions of a single variable presented in this book. This is possible because of the convention that $C(n, k) = 0$ if $k < 0$ or $k > n$. For the multivariable versions, the domain will be the tuples $(k_1, \ldots, k_t)$, where $\sum k_i = n$. This is possible because of the convention that $M(k_1, \ldots, k_t) = 0$ if any of the $k_i < 0$. The support of a function is all values for which the function is non-zero. For example, while the binomial distribution has a domain of all integers, its support is the integers between zero and the number of trials.

5.4 MULTINOMIAL DISTRIBUTION

Unlike *Zombicide: Black Plague* (p.87), where only one result was tracked (a hit or a miss), other games require tracking more than one result. In *King of Tokyo*, players take the role of rampaging monsters battling each other in Tokyo. One route to victory is accumulating 20 victory points. Players will do this by rolling six six-sided dice on their turn. Each die has the following sides: a one, a two, a three, a heart, a claw, and a lightning bolt. Sets of numbered sides allow the player to earn victory points, the heart allows the player's monster to heal, the claw allows the player's monster to attack another monster, and the lightning bolt allows the player to collect energy, which they can spend to increase their monster's power. Here, the focus will be on the victory points that a player can accumulate. A set of three dice with the same number will earn the player the number of victory points shown on the dice. For example, a set of three twos will earn the player two victory points. Once a set of three is achieved, each additional die showing that side will earn an additional victory point. So, for example, a set of four twos will earn the player $2+1 = 3$ victory points (the first two from the set of three and the third point from the fourth two). Like *Yahtzee*, players may re-roll dice up to two times (for three total rolls), keeping any number of dice from one roll to the next. Assume a player has rolled two ones, one three, and three sides with symbols. How many victory points should they expect if they retain all three numbers and re-roll the remaining three dice once?

Tracking only the ones or tracking only the threes would be a binomial distribution, but here, both of these values must be tracked. Furthermore, these sides cannot be tracked separately as they are not independent (rolling a one on a die precludes rolling a three on that die). Ignoring the case of rolling three twos for the moment, there are three relevant categories of dice sides: a one, a three, and everything else (denoted by $\star$). With three categories and three dice, there are $3^3 = 27$ possible rolls. Only rolls that lead to positive victory points will contribute to the expected value, so the probabilities of rolls leading to zero victory points can be ignored. The order in which the results were rolled will also be ignored in the first column of the unshaded rows of Table 5.1.

Every one is rolled with a probability of $1/6$, every three is rolled with a probability of $1/6$, and every $\star$ is rolled with a probability of $4/6$. This information allows the second column of the unshaded rows to be filled in. All three probabilities are included in each expression even if the exponent is zero, as this will help clarify the pattern (and make it easier to write the final result with a summation).

Finally, it is necessary to determine the number of leaves in the probability tree that correspond to each outcome. For example, the number of times outcome $1\star\star$ occurs is the number of permutations of one one, zero threes, and two stars. This is precisely the multinomial coefficient $M(1, 0, 2)$. This fills in the third column of the unshaded rows. The binomial coefficients could be used in this case (since there are at most two non-zero values in each multinomial expression). However, writing this with a multinomial coefficient again makes the final summation and the generalizations easier to see.

The one case that still needs to be handled is rolling three twos. There are two ways to handle this situation. The first method would be tracking twos in addition to ones and threes. While this can be done and might be required in more complicated situations, in this situation, there is only one way to score with twos (to roll a set of three twos on the re-roll), and tracking twos is not worth the added complication. Instead, this exceptional case is handled separately from the main calculation (and is the shaded row of the table). Summing the last column yields a value of approximately 0.74.

This introduces the multinomial distribution whose PMF is presented in the Multinomial Distribution Properties on page 99. While these distributions are a natural extension of the binomial distributions, the formal description gets more complicated as it must account for any number of tracked outcomes. In practice, the formula is simpler to use than it appears here, as t, n, and the p_i are fixed from the start of the process.

Table 5.1: Outcomes in *King of Tokyo*.

Outcome	Probability	Orderings	Expected Victory Points
1★★	$(1/6)^1 (1/6)^0 (4/6)^2$	$\binom{3}{1,0,2}$	$1 \times \binom{3}{1,0,2} (1/6)^1 (1/6)^0 (4/6)^2$
13★	$(1/6)^1(1/6)^1(4/6)^1$	$\binom{3}{1,1,1}$	$1 \times \binom{3}{1,1,1}(1/6)^1(1/6)^1(4/6)^1$
11★	$(1/6)^2 (1/6)^0 (4/6)^1$	$\binom{3}{2,0,1}$	$2 \times \binom{3}{2,0,1} (1/6)^2 (1/6)^0 (4/6)^1$
113	$(1/6)^2(1/6)^1(4/6)^0$	$\binom{3}{2,1,0}$	$2 \times \binom{3}{2,1,0}(1/6)^2(1/6)^1(4/6)^0$
111	$(1/6)^3 (1/6)^0 (4/6)^0$	$\binom{3}{3,0,0}$	$3 \times \binom{3}{2,0,1} (1/6)^2 (1/6)^0 (4/6)^1$
33★	$(1/6)^0 (1/6)^2 (4/6)^1$	$\binom{3}{0,2,1}$	$3 \times \binom{3}{0,2,1} (1/6)^0 (1/6)^2 (4/6)^1$
333	$(1/6)^0 (1/6)^3 (4/6)^0$	$\binom{3}{0,3,0}$	$4 \times \binom{3}{0,3,0} (1/6)^0 (1/6)^3 (4/6)^0$
133	$(1/6)^1 (1/6)^2 (4/6)^0$	$\binom{3}{1,2,0}$	$4 \times \binom{3}{1,2,0} (1/6)^1 (1/6)^2 (4/6)^0$
222	$(1/6)^3$	1	$2 \times (1/6)^3$

Multinomial Distribution Properties: Consider a series of n independent trials each with t possible outcomes where the ith outcome has probability p_i (so $\sum_{i=1}^{t} p_i = 1$) and let X_i be the number of times outcome i occurs. The probability that $X_1 = k_1$, $X_2 = k_2$, ..., $X_t = k_t$ (so $\sum_{i=1}^{t} k_i = n$) is

$$P[X_1 = k_1, \ldots, X_t = k_t] = \binom{n}{k_1, k_2, \ldots, k_t} p_1^{k_1} p_2^{k_2} \cdots p_t^{k_t}.$$

In *Black Orchestra*, players take on the role of German citizens attempting to assassinate Adolph Hitler during the events of World War II. When the characters have collected the required items for a plot, they will roll dice to determine the success or failure of their mission using a custom six-sided die with the following sides: a one, a two, a three, an eagle, and two sides with a target symbol. After rolling the die, the players first check the number of eagles. If this exceeds a specified threshold (which will change during the game), the plot will be uncovered, and the characters will be imprisoned. If the plot is not uncovered, the players will check the number of targets, and the plot will succeed if this exceeds a specified threshold (which also changes during the game). For illustration purposes, assume that seven dice will be rolled and that the characters will be detected if there are two or more eagles and will succeed if there are four or more targets. What is the probability of success?

The binomial distribution indicates that the probability of rolling one or fewer eagles is approximately 67%:

$$\underbrace{\binom{7}{0}\left(\frac{1}{6}\right)^0\left(\frac{5}{6}\right)^7}_{\text{0 Eagles}} + \underbrace{\binom{7}{1}\left(\frac{1}{6}\right)^1\left(\frac{5}{6}\right)^6}_{\text{1 Eagle}} \approx 0.67.$$

Similarly, the probability of rolling four or more targets is approximately 17%:

$$\binom{7}{4}\left(\frac{2}{6}\right)^4\left(\frac{4}{6}\right)^3 + \cdots + \binom{7}{7}\left(\frac{2}{6}\right)^7\left(\frac{4}{6}\right)^0 \approx 0.17.$$

Again, multiplying these two probabilities does not correctly determine the probability of success. The problem is that the two events are not independent: knowing how many eagles are rolled will affect the chance of how many targets are rolled. For example, knowing that seven eagles were rolled, the chance of rolling four or more targets would be 0%, much less than the 17% calculated above.

The first step is to break this compound event into the seven outcomes associated with success: no eagles with four to seven targets and one eagle with four to six targets. Here, the probability of one eagle and five targets will be calculated, and then the general formula will be developed. Categorize each roll as one of three types: eagle, target, or number. There is a $1/6$ probability of rolling an eagle, a $2/6$ probability of rolling a target, and a $3/6$ probability of rolling a number. So, any leaf which results in one eagle, five targets, and one number has the probability accumulated along the branches of $(1/6)^1 (2/6)^5 (3/6)^1$. To determine the number of leaves associated with this outcome, use the multinomial coefficient $M(1,5,1)$. So,

$$P[1 \text{ Eagle and 5 Targets}] = \binom{7}{1,5,1}\left(\frac{1}{6}\right)^1\left(\frac{2}{6}\right)^5\left(\frac{3}{6}\right)^1.$$

Alternatively, one can arrive at this formula using the conditional probability that five targets will be rolled, given that exactly one eagle has been rolled, which leads to the same expression. Returning to *Black Orchestra*, the probability of success is the sum of the 7 unshaded possibilities in Table 5.2.

Let i be the number of Eagles, j be the number of Targets, and k be the number of numbers, then each of the seven unshaded entries in the table can be written as $M(i,j,k) (1/6)^i (2/6)^j (3/6)^k$, where $i+j+k=7$. In Section 1.3, it was noted that $M(i,j,k) = 0$ if any of i, j, or k were negative and that can be used here to fill the remaining table entry as well. In the case where $i=1$ and $j=7$, then $k=-1$ as $i+j+k=7$ and so the shaded entry, zero, can also be written as $M(i,j,k) (1/6)^i (2/6)^j (3/6)^k$. This makes it easier to compute values like this in computer algebra systems. So there is approximately a 15%

Table 5.2: Possible Successes in *Black Orchestra*.

Targets	Eagles 0	Eagles 1
	0	**1**
4	$\binom{7}{0,4,3}(1/6)^0(2/6)^4(3/6)^3$	$\binom{7}{1,4,2}(1/6)^1(2/6)^4(3/6)^2$
5	$\binom{7}{0,5,2}(1/6)^0(2/6)^5(3/6)^2$	$\binom{7}{1,5,1}(1/6)^1(2/6)^5(3/6)^1$
6	$\binom{7}{0,6,1}(1/6)^0(2/6)^6(3/6)^1$	$\binom{7}{1,6,0}(1/6)^1(2/6)^6(3/6)^0$
7	$\binom{7}{0,7,0}(1/6)^0(2/6)^7(3/6)^0$	0

chance of success:

$$\sum_{\substack{0\leq i\leq 1 \\ 4\leq j\leq 7 \\ i+j+k=7}} \binom{7}{i,j,k}\left(\frac{1}{6}\right)^i\left(\frac{2}{6}\right)^j\left(\frac{3}{6}\right)^k \approx 0.15.$$

Some consistency checks can be done when computing sums for events from a multinomial distribution. The sum of the probabilities in each term ($1/6$, $2/6$, and $3/6$) should be one. The exponents of the probabilities in each term (i, j, k) should match the lower indices of the multinomial coefficient. Including extraneous indices (for example, not reducing $M(0, 4, 3)$ to $C(7, 4)$), or factors (for example, including the factor $(1/6)^0 = 1$) is often worthwhile to make it easier to verify the consistency of the formula and to make general patterns clear. It can also make entering the expression as a single sum on a computer algebra system easier.

5.5 HYPERGEOMETRIC DISTRIBUTION

In *Flamme Rouge* (p.13), players move by drawing cards from a deck instead of rolling dice. The multinomial distribution (and the binomial distribution) cannot model this type of randomness because the trials are not independent here. Drawing a nine as the first card decreases the number of nines remaining in the deck and reduces the chance of drawing another nine. To reduce the situation's complexity, assume the cyclist's deck has been reduced to six cards: two twos, one four, one five, and two nines. What is the probability that they will draw precisely one nine in this hand (which means they will draw the second nine on their next turn)?

This is now a counting problem. There are two cards in one category (the nines) and four cards in a second category (the twos, fours, and fives). There are $C(2, 1)$ ways to select one card from the first category and $C(4, 2)$ ways to select a card from the second category. Because the nines and the other cards are disjoint, the Product Rule calculates that there are $C(2, 1)C(4, 2)$ total ways to select exactly one nine in a hand of three cards. Divide this by the $C(6, 3)$ ways to select three cards from the six in the deck to arrive at the probability of approximately 60%:

$$P[\text{exactly one } 9] = \frac{\binom{2}{1}\binom{4}{2}}{\binom{6}{3}} \approx 0.60.$$

This is the hypergeometric distribution and is used when one is selecting *without* replacement. The hypergeometric distribution uses terminology similar to that of the binomial distribution. In the example above, drawing a card constitutes a trial, resulting in either a success or a failure. The objects being drawn from is the population. The general formula for the hypergeometric distribution is calculated the same way as in the *Flamme Rouge* example, and the expected value can be calculated the same way as for the binomial distribution. These results are collected in the Hypergeometric Distribution Properties.

Hypergeometric Distribution Properties: The probability of k observed successes for an n-trial hypergeometric distribution with N outcomes (where $n \leq N$), of which K are successes is

$$P[\text{exactly } k \text{ successes}] = \frac{\binom{K}{k}\binom{N-K}{n-k}}{\binom{N}{n}}.$$

The expected value of this distribution is $\mu = n \times K/N$.

Like the Binomial Distribution, the Hypergeometric Distribution has a multivariable version, called the multivariable hypergeometric distribution. In the multivariable case, the distribution tracks multiple outcomes. As with the Multinomial Distribution Properties, the notation becomes complicated to handle all cases. The calculation is easier in practice, where t, N, and the K_i are all fixed from the start of the process.

Multivariable Hypergeometric Distribution Properties: If there is a series of n trials with t possible outcomes being selected without replacement from an initial collection of size N where there are K_i outcomes of type i (so $\sum_{i=1}^{t} K_i = N$). Let X_i be the number of times outcome i occurs , the probability that $X_1 = k_1$, $X_2 = k_2$, ..., $X_t = k_t$ (so $\sum_{i=1}^{t} k_i = n$) is

$$P[X_1 = k_1, X_2 = k_2, \ldots, X_t = k_t] = \frac{\binom{K_1}{k_1}\binom{K_2}{k_2}\cdots\binom{K_t}{k_t}}{\binom{N}{n}}.$$

Also, like the multinomial distribution, some consistency checks can be done. For instance, the upper indices in the numerator should sum to the upper index in the denominator, and the lower indices in the numerator should sum to the lower index in the denominator.

If N is "large" relative to n, the values obtained by hypergeometric distribution can be approximated by the values obtained from the binomial distribution (or a multinomial distribution for the multivariable case), where $p = K/N$. However, in tabletop gaming, a computer can often compute the hypergeometric values as efficiently as the approximation. Section 5.6 discusses the practical difference between these two distributions.

5.6 COMPARING DISTRIBUTIONS

In *Oathsworn: Into the Deep Wood* (p.86), players can roll dice *or* draw cards to determine damage. The lowest die in the game is a white die, which has two blank sides, two sides with a one, one side with a two, and one side with an exploding two (rolling this side causes two damage and allows a player to roll a bonus white die). Players drawing a card from a newly shuffled white deck will be drawing from a deck of 18 cards: six blank cards, six cards with a one, three cards with a two, and three cards with an exploding two. From a full deck, a player has a $2/6$ chance of drawing a blank card, a $2/6$ chance of drawing a card with a one, a $1/6$ chance of drawing a card with a two, and a $1/6$ chance of drawing a card with an exploding two, which matches the probabilities of rolling these symbols on a white die. All examples in this section assume the deck contains all 18 cards at the start of the attack.

The steps in an attack will be described as if the player chose to draw cards instead of roll dice. However, in any place where the steps refer to drawing cards, a player may choose to roll dice instead. A player first decides how many cards (up to ten) they wish to draw and makes their initial draw. They score a hit if their initial draw contains one or fewer blank cards. They then draw any bonus cards resulting from exploding twos. They draw one bonus card for every exploding two as an initial or bonus card. Drawing blank cards on bonus draws will not cause the character to miss; the check for missing only occurs on the initial draw. What is the probability of hitting if the player rolls three white dice? What is the probability of hitting if the player draws three white cards?

If rolling three dice, a player must roll either zero blank sides or one blank side. Both of these probabilities can be computed using the binomial distribution, where $n = 3$, $p = 2/6$, and $k = 0$ or $k = 1$. Adding these values arrives at a 74% probability of hitting:

$$\underbrace{\binom{3}{0}\left(\frac{2}{6}\right)^0\left(\frac{4}{6}\right)^3}_{k=0} + \underbrace{\binom{3}{1}\left(\frac{2}{6}\right)^1\left(\frac{4}{6}\right)^2}_{k=1} = \frac{20}{27} \approx 0.74.$$

Table 5.3: Hit Probabilities in *Oathsworn: Into the Deep Wood*.

Method	Number of Dice or Cards								
	2	3	4	5	6	7	8	9	10
Dice	89%	74%	59%	46%	35%	26%	20%	14%	10%
Cards	90%	75%	59%	44%	31%	20%	12%	7%	3%

If drawing three cards, a player must draw either zero blank cards or one blank card. Both of these probabilities can be computed using the hypergeometric distribution, where $N = 18$, $K = 6$, $n = 3$, and $k = 0$ or $k = 1$. Adding these values arrives at a 75% chance of hitting:

$$\underbrace{\frac{\binom{6}{0}\binom{12}{3}}{\binom{18}{3}}}_{k=0} + \underbrace{\frac{\binom{6}{1}\binom{12}{2}}{\binom{18}{3}}}_{k=1} = \frac{77}{102} \approx 0.75.$$

So drawing three cards and rolling three dice have hit probabilities very close to each other, a difference of only thirteen hits for every 918 attacks. Doing this same calculation for the number of cards drawn or dice rolled between two (the fewest that could result in a miss) and ten (the most a player may use) yields Table 5.3. This shows that dice become more effective than cards when five or more are used.

Once a player has hit, damage is calculated. Unfortunately, achieving a hit and the damage inflicted are not independent. With a hit, one knows that, at most, one card is blank, which means that the sum on the cards of a successful attack will generally be higher than the sum on the cards of an unsuccessful attack. How much damage is expected if a player rolls three white dice in an attack? How much damage is expected if a player draws three white cards in an attack?

For dice, the situation is relatively straightforward. The expected value of an exploding two can be computed in the same way it was computed for *Zombicide: Black Plague* (p.87) in Section 5.2 to arrive at $\mu = {}^6\!/_5$. A helper function d is used to determine the amount of damage a hit will cause if i blank cards, j cards with ones, k cards with twos, and ℓ cards with exploding twos are drawn. The result can be computed by summing over all the possibilities where $i \leq 1$ (a hit) and $i + j + k + \ell = 3$ (as there are only three dice). Because the multinomial coefficient $M(i, j, k, \ell)$ is zero if any of i, j, k, or ℓ is a negative number, the bounds on j, k, and ℓ do not need to be specified here. The result is an average damage of approximately 3.2:

$$d(i, j, k, \ell) = 0 \times i + 1 \times j + 2 \times k + (2 + \mu) \times \ell,$$

$$\sum_{\substack{i \leq 1 \\ i+j+k+\ell=3}} d(i, j, k, \ell) \times \binom{3}{i, j, k, \ell} \left(\frac{2}{6}\right)^i \left(\frac{2}{6}\right)^j \left(\frac{1}{6}\right)^k \left(\frac{1}{6}\right)^\ell \approx 3.2.$$

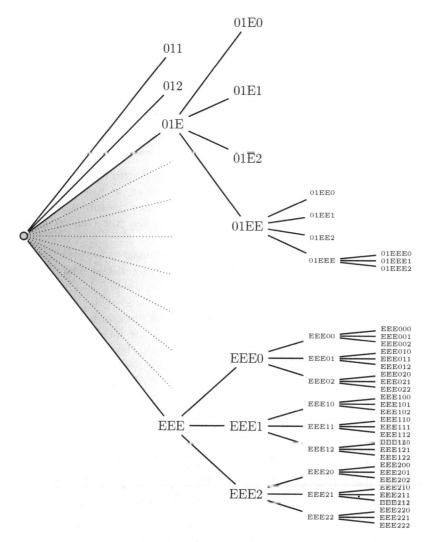

Figure 5.3: Probability Tree for *Oathsworn: Into the Deep Wood*.

Note that this isn't equal to 3.6, which is three times the expected value of a single die roll. This is because when computing $3 \times E[\text{Damage with 1 Die}]$, the sequence blank-blank-two contributes two damage to the numerator of the expected value. However, the sequence blank-blank-two does not contribute to $E[\text{Damage with 3 Dice}]$ because the two blanks cause the attack to miss.

Computing the expected damage while using the deck is more complicated. In this case, a probability tree for events leading to non-zero damage can be constructed. Figure 5.3 shows some of this tree, where a "0" indicates a blank card and an "E" indicates a card with an exploding two. The edges from the

Table 5.4: Expected Damage in *Oathsworn: Into the Deep Wood*.

Method	Number of Dice or Cards								
	2	3	4	5	6	7	8	9	10
Dice	2.40	3.20	3.56	3.56	3.32	2.95	2.53	2.11	1.72
Deck of 90	2.39	3.18	3.52	3.49	3.21	2.80	2.34	1.88	1.47
Deck of 18	2.35	3.13	3.40	3.22	2.73	2.09	1.44	0.88	0.47

root can be computed using the hypergeometric distribution, as there will always be an initial draw of three cards, and there are at least three of each card in the deck. However, the depth of the tree after the initial three draws depends on what has been drawn. For example, the tree ends at outcome "011" but will continue after "01E."

Direct calculation of the expected amount of damage is beyond hand computation, but this tree can be transversed by a depth-first search using a computer algebra system. The results are shown in Table 5.4. The table also includes the results assuming the damage deck consisted of 90 cards instead of the standard 18, illustrating that the binomial distribution becomes a better fit for the hypergeometric distribution as the deck size increases.

Given this information, asking why anyone would choose cards over dice is reasonable. However, there are advantages to using cards that would need to be included in a complete analysis. Unlike dice, cards guarantee a minimum number of successes as one cannot draw more than 6 blank cards in a deck of 18 cards. Players can also track the number and type of cards remaining in the decks. For example, players can safely draw more cards after many poor cards have already been drawn. In the extreme case, if the first five cards are all blank, there is no reason not to draw the maximum ten cards on the next attack (as, at most, one blank card remains in the deck). On the other hand, if the players have already drawn many of the better cards and the deck is now filled with blank cards, the player can use an ability in the game to reshuffle the discard pile back into the deck, essentially getting the earlier good draws without having to pay for them with subsequent poor draws.

5.7 MARKOV PROCESSES

In many games involving combat, players may choose to continue the combat or retreat after each round of combat. One aspect of this situation is that the state of the board determines the outcome of each round of combat without regard to how the board reached that state. A process for which this is true is known as a Markov process or Markov chain.[2]

---

[2]Markov processes are named after Andrey Andreyevich Markov, a twentieth-century Russian mathematician known for his work in stochastic processes.

In *Star Wars: Rebellion*, one player takes the role of the Empire while the other takes the role of the Rebellion. As expected in a Star Wars game, combat between spaceships is a part of the gameplay. The Empire has tie-fighters at its command, and the Rebellion has Y-wings. Both ships will be destroyed with a single hit from the opponent. The tie-fighter has a one-in-two chance of hitting the Y-wing, while the Y-wing only has a one-in-six chance of hitting the tie-fighter. Assume neither ship will retreat, and combat will continue until one or both ships are destroyed. It seems clear that the tie-fighter will likely win the encounter, but what is the probability of a tie fighter victory, and how long will the combat take to resolve?

This problem can be solved using the same technique from the exploding dice example. Let p be the probability that only the Y-wing is destroyed. Two possible outcomes of the first round could lead to this outcome. The Y-wing could be destroyed in the first round of combat, or both ships could survive, and the Y-wing's destruction occurs later. The key is that if both ships survive into the next round of combat, the situation at the start of the next round is the same as at the start of the current round of combat. If both ships survive, the probability that only the Y-wing is eventually destroyed is p again. This produces the equation

$$p = P[\text{destroyed first round}]$$
$$+ P[\text{both survived}] \, P[\text{destroyed later} \mid \text{both survived}]$$
$$= \frac{5}{12} + \frac{5}{12}p.$$

Solving determines that if this value exists, it must be $p = 5/7$. However, the goal is to develop a general solution for more complicated cases.

The first step toward the solution is to set up a probability tree as shown in Figure 5.4, where the labels indicate which ship was destroyed. The tree is infinite because the battle continues as long as neither ship is destroyed. However, if neither ship is destroyed in an attack, the state of the game is unchanged. So, instead of tracking a series of no-hit results, a loop is drawn from the initial vertex to itself. As a result, the graph is no longer a tree, but it is directed, so arrowheads will be necessary to track the directions on the edges. This creates the graph in Figure 5.5. The initial vertex remains shaded to make it easy to identify, and the terminal vertices have a double border around them, indicating a conclusion of the process (with the destruction of one or both ships).

A diagram like this represents a finite state machine, showing a finite number of states (the vertices) and the transitions between the available states. In this case, the transitions are labeled with the probability that they will occur. At this point, the techniques for computations associated with traversing a graph can be leveraged. A weighted adjacency matrix from Section 4.2 can be used to track the probability of arriving at any game state. The states are ordered as follows: the initial state (where both ships are still fighting),

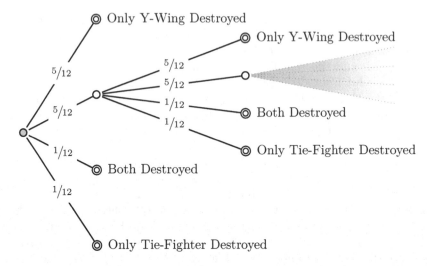

Figure 5.4: Probability Tree for Y-Wing versus Tie-Fighter in *Star Wars: Rebellion*.

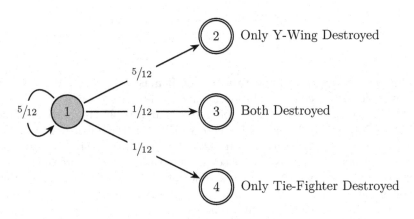

Figure 5.5: Transition Graph for Y-Wing versus Tie-Fighter in *Star Wars: Rebellion*.

the state where only Y-wing is destroyed, the state where both ships are destroyed, and the state where only tie-fighter destroyed. The first column of the matrix will be $\langle 5/12, 5/12, 1/12, 1/12 \rangle$, where the notation $\langle \ldots \rangle$ indicates that the values represent a column in the matrix. Once the state is one of the final states (with one or both of the ships destroyed), the state will not change. So the second column will be $\langle 0, 1, 0, 0 \rangle$, the third column will be $\langle 0, 0, 1, 0 \rangle$, and the fourth column will be $\langle 0,0,0,1 \rangle$. Putting these together, the transition matrix is

$$\mathbf{P} = \begin{pmatrix} \frac{5}{12} & 0 & 0 & 0 \\ \frac{5}{12} & 1 & 0 & 0 \\ \frac{1}{12} & 0 & 1 & 0 \\ \frac{1}{12} & 0 & 0 & 1 \end{pmatrix}.$$

It should be noted that there are different conventions for probability transition matrices. The convention followed here (where (i,j) entry corresponds to transitioning from j to i) was chosen here because it matches the earlier work on transition matrices. Within the probability community, the convention that the (i,j) entry corresponds to the transition from i to j is more common. To determine which convention is being used, sum the rows and columns. In the convention used here, all columns will sum to one. For the other convention, all rows will sum to one.

Because sequential rounds result in multiplying the probabilities, the square of the matrix will represent probabilities associated with two rounds of the battle. In this case,

$$\mathbf{P}^2 = \begin{pmatrix} \frac{25}{144} & 0 & 0 & 0 \\ \frac{85}{144} & 1 & 0 & 0 \\ \frac{17}{144} & 0 & 1 & 0 \\ \frac{17}{144} & 0 & 0 & 1 \end{pmatrix}.$$

This matrix contains the probabilities of transitioning from one state to another in two steps. For example, it is possible to transition from the initial state to the state where only the Y-wing is destroyed in two ways: only the Y-wing was destroyed in the first round, or both ships survived the first round, with only the Y-wing destroyed in the second round. The probability of only the Y-wing being destroyed in the first round is $5/12 = 60/144$ while the probability of surviving the first round, only to be destroyed in the second round is $(5/12)(5/12) = 25/144$. The sum of these is $85/144$, matching the value in the matrix.

In this case, the only column of interest is the first column (from starting in the initial state). The notation $\mathbf{P}[i, \cdots]$ will be used to represent the ith row of a matrix and $\mathbf{M}[:, j]$ will be used to represent the jth column of a matrix,

$$\mathbf{P}^2[:, 1] = \left\langle \frac{25}{144}, \frac{85}{144}, \frac{17}{144}, \frac{17}{144} \right\rangle.$$

It should be noted that remaining in the initial state becomes increasingly less probable with each round, having a probability of $(5/12)^n$ in the nth round, and as n grows without bound, this probability drops to zero. After ten rounds of combat, there is a 71% chance that only the Y-wing will be destroyed, a 14% chance only the tie-fighter will be destroyed, and a 14% chance that both ships will be destroyed:

$$\mathbf{P}^{10}[:,1] \approx \langle 0.00016, 0.71417, 0.14284, 0.14284 \rangle .$$

As n grows, the probability of the Y-wing being destroyed will approach $5/7 \approx 71.43\%$, which is very close to the value determined for $n = 10$:

$$\left(\lim_{n \to \infty} \mathbf{P}^n \right)[:,1] = \left\langle 0, \frac{5}{7}, \frac{1}{7}, \frac{1}{7} \right\rangle .$$

To determine the expected number of rounds in the battle, the formula for the expected value can be used, where there is a $7/12$ probability of the battle concluding and a $5/12$ probability of the battle continuing to the next round,

$$\mu = 1 \left(\frac{7}{12} \right) + 2 \left(\frac{5}{12} \right) \left(\frac{7}{12} \right) + \cdots + k \left(\frac{5}{12} \right)^{k-1} \left(\frac{7}{12} \right) + \cdots$$

The exploding dice technique cannot be used on this sum due to the factor of k in each term, but the sum can be calculated using calculus, obtaining $\mu = 12/7$, so the battle is expected to last between 1 and 2 rounds.

However, this Markov process is known as an absorbing Markov process since some states only transition to themselves. Once one of the ships is destroyed, the state will not change. As a result, the Markov Absorption Property on page 110 can be used.

Markov Absorption Property: Given an absorbing Markov process where there is a positive transition probability from every state to at least one absorbing state, then the states can be chosen so that the process' weighted transition matrix, $\mathbf{P}$, is in the form

$$\mathbf{P} = \left(\begin{array}{c|c} \mathbf{Q} & \mathbf{0} \\ \hline \mathbf{R} & \mathbf{I} \end{array} \right),$$

for some matrices $\mathbf{Q}$ and $\mathbf{R}$, where $\mathbf{0}$ is a matrix of zeros and $\mathbf{I}$ is an identity matrix. Setting $\mathbf{N} = \sum_{k=0}^{\infty} \mathbf{Q}^k$,

$$\lim_{k \to \infty} \mathbf{P}^k = \left(\begin{array}{c|c} \mathbf{0} & \mathbf{0} \\ \hline \mathbf{RN} & \mathbf{I} \end{array} \right)$$

The expected number of steps before being absorbed from state i is equal to the value in the ith column of the matrix

$$\mathbf{1N},$$

where $\mathbf{1}$ is a row matrix consisting of all ones.

In the specific example of the Y-wing and tie-fighter,

$$
\left(\begin{array}{c|c}
\mathbf{Q} & \mathbf{0} \\
\hline
\mathbf{R} & \mathbf{I}
\end{array}\right)
=
\left(\begin{array}{c|ccc}
\frac{5}{12} & 0 & 0 & 0 \\
\hline
\frac{5}{12} & 1 & 0 & 0 \\
\frac{1}{12} & 0 & 1 & 0 \\
\frac{1}{12} & 0 & 0 & 1
\end{array}\right)
$$

Using a computer algebra system to compute $\mathbf{N} = \left(\frac{12}{7}\right)$, the limit can be derived as

$$
\lim_{k \to \infty} \mathbf{P}^k =
\left(\begin{array}{c|c}
\mathbf{0} & \mathbf{0} \\
\hline
\mathbf{RN} & \mathbf{I}
\end{array}\right)
=
\left(\begin{array}{c|ccc}
0 & 0 & 0 & 0 \\
\hline
\frac{5}{7} & 1 & 0 & 0 \\
\frac{1}{7} & 0 & 1 & 0 \\
\frac{1}{7} & 0 & 0 & 1
\end{array}\right).
$$

The expected number of steps for the battle to end is equal to the value in the first column of $\mathbf{1N} = \left(\frac{12}{7}\right)$, which also matches our previous result that average number of turns to transition from the initial state to an absorbing state is $\frac{12}{7}$.

A More Interesting Space Battle

A more exciting application arises in a larger space battle from *Star Wars: Rebellion*: an attack of a Y-wing and Corellian corvette on a star destroyer. The Y-wing can take one hit, the corvette can take two, and the destroyer can take four hits. Assume the destroyer will disengage and retreat from battle if it takes two hits and any rebel ship remains. There are 14 relevant states in this situation as listed in Table 5.5. The shaded states are absorbing states, indicating an end to the battle.

Figure 5.6 gives all the states in the finite state machine graph and some transition arrows. The graph is best thought of as a three-dimensional grid. The upper layer (States 1–6 shown on the left) consists of those states, where the Y-wing is undamaged, and the lower layer (States 7–11 shown on the right) consists of those states where the Y-wing has been destroyed. Moving to the right indicates damage to the corvette, while moving down indicates damage to the destroyer. States 12 and 13 occur in both layers, indicating a Rebel Victory. State 14, the destruction of all ships only occurs on the bottom layer. Arrows that loop back to the same state, arrows that result from chaining the arrows shown, and arrows from the graph's upper layer to the graph's lower layer are omitted.

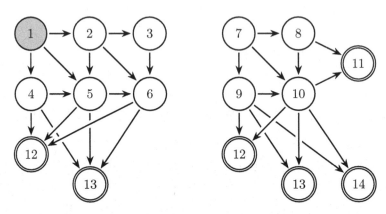

Figure 5.6: Transition Graph for an Exciting Space Battle in *Star Wars: Rebellion.*

Both the corvette and Y-wing will target the destroyer. The destroyer's priority from highest to lowest will be to destroy the corvette, destroy the Y-wing, and damage the corvette. The probability of success for each of these was calculated using a computer algebra system. This information was used to construct the transition matrix shown in Figure 5.7.

Table 5.5: Combat States for an Exciting Space Battle in *Star Wars: Rebellion.*

Damage		Y-wing	
Corvette	Destroyer	Undamaged	Destroyed
0	0	1	7
1	0	2	8
2	0	3	11
0	1	4	9
1	1	5	10
2	1	6	11
Empire Retreats		12	12
Destroyer Destroyed		13	13
All Ships Destroyed		–	14

0.03	0.	0.	0.	0.	0.	0.	0.	0.	0.	0.	0.	0.	0.
0.03	0.03	0.	0.	0.	0.	0.	0.	0.	0.	0.	0.	0.	0.
0.04	0.07	0.17	0.	0.	0.	0.	0.	0.	0.	0.	0.	0.	0.
0.06	0.	0.	0.03	0.	0.	0.	0.	0.	0.	0.	0.	0.	0.
0.08	0.06	0.	0.03	0.03	0.	0.	0.	0.	0.	0.	0.	0.	0.
0.1	0.16	0.17	0.04	0.07	0.17	0.	0.	0.	0.	0.	0.	0.	0.
0.04	0.	0.	0.	0.	0.	0.09	0.	0.	0.	0.	0.	0.	0.
0.03	0.02	0.	0.	0.	0.	0.19	0.09	0.	0.	0.	0.	0.	0.
0.1	0.	0.	0.04	0.	0.	0.1	0.	0.09	0.	0.	0.	0.	0.
0.08	0.04	0.	0.03	0.02	0.	0.23	0.1	0.19	0.09	0.	0.	0.	0.
0.12	0.44	0.65	0.12	0.43	0.65	0.33	0.79	0.33	0.79	1.	0.	0.	0.
0.29	0.19	0.	0.66	0.42	0.17	0.06	0.02	0.39	0.12	0.	1.	0.	0.
0.	0.	0.	0.04	0.02	0.	0.	0.	0.	0.	0.	0.	1.	0.
0.	0.	0.	0.01	0.02	0.	0.	0.	0.	0.	0.	0.	0.	1.

Figure 5.7: Transition Matrix for an Exciting Space Battle in *Star Wars: Rebellion*.

The matrix $\mathbf{N}$ can be computed by a computer algebra system and is

$$
\mathbf{N} =
\begin{pmatrix}
1.03 & 0. & 0. & 0. & 0. & 0. & 0. & 0. & 0. & 0. \\
0.04 & 1.03 & 0. & 0. & 0. & 0. & 0. & 0. & 0. & 0. \\
0.06 & 0.09 & 1.21 & 0. & 0. & 0. & 0. & 0. & 0. & 0. \\
0.06 & 0. & 0. & 1.03 & 0. & 0. & 0. & 0. & 0. & 0. \\
0.08 & 0.06 & 0. & 0.04 & 1.03 & 0. & 0. & 0. & 0. & 0. \\
0.15 & 0.22 & 0.25 & 0.06 & 0.09 & 1.21 & 0. & 0. & 0. & 0. \\
0.05 & 0. & 0. & 0. & 0. & 0. & 1.10 & 0. & 0. & 0. \\
0.05 & 0.02 & 0. & 0. & 0. & 0. & 0.23 & 1.10 & 0. & 0. \\
0.12 & 0. & 0. & 0.05 & 0. & 0. & 0.12 & 0. & 1.10 & 0. \\
0.13 & 0.05 & 0. & 0.05 & 0.02 & 0. & 0.33 & 0.12 & 0.23 & 1.10 \\
\end{pmatrix}.
$$

The probability of outcomes can be determined by the matrix $\mathbf{RN}$,

$$
\mathbf{RN} =
\begin{pmatrix}
0.52 & 0.74 & 0.96 & 0.23 & 0.51 & 0.79 & 0.85 & 0.97 & 0.55 & 0.87 \\
0.47 & 0.26 & 0.04 & 0.72 & 0.45 & 0.21 & 0.15 & 0.03 & 0.45 & 0.13 \\
0. & 0. & 0. & 0.04 & 0.02 & 0. & 0. & 0. & 0. & 0. \\
0. & 0. & 0. & 0.01 & 0.02 & 0. & 0. & 0. & 0. & 0. \\
\end{pmatrix}.
$$

The columns of $\mathbf{RN}$ are associated with non-absorbing states, and the rows are associated with the absorbing states. Starting in the initial state (State 1), there is a 52% chance of an Empire victory (State 11) and a 47% chance of an Empire Retreat (State 12). If the battle had started with one damage on the Destroyer (State 4), there would be a 23% chance of an Empire victory (State 11), a 72% chance of an Empire retreat (State 12), a 4% chance of the destruction of the Destroyer (State 13), and a 1% chance of the destruction of all ships. If the battle had started without the Y-wing (State 7), there would have been an 85% chance of an Empire victory (State 11) and a 15% chance of an Empire retreat (State 12). One effective way to organize this information (and to quickly check that the numbers trend in the expected way) is to label

—— Y-Wing Undamaged ——   —— Y-Wing Destroyed ——

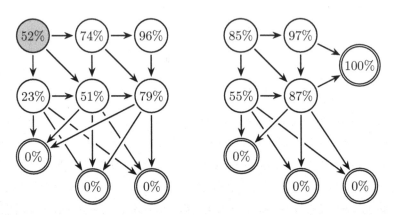

Figure 5.8: Probability of Empire Victory for an Exciting Space Battle in *Star Wars: Rebellion*.

each vertex with the probability of an Empire victory (State 11) from that state. This is done in Figure 5.8.

Computing the matrix **1N**,

$$\mathbf{1N} = \begin{pmatrix} 1.8 & 1.5 & 1.5 & 1.2 & 1.1 & 1.2 & 1.8 & 1.2 & 1.3 & 1.1 \end{pmatrix}.$$

This indicates that the exciting battle will be over quickly, with the expected number of rounds between one and two for any starting state. The highest expected value is from the initial state, which is not unexpected and is 1.8 rounds.

Markov processes are also valuable for predicting game locations, where card draws or dice rolls dictate movement. However, this roll-and-move mechanism is not popular among the hobby game community because of the lack of player agency. Games such as *Snakes and Ladders*, *Chutes and Ladders*, and *Candy Land* all have absorbing states and can be analyzed using the Markov Absorption Property and references to these analyses can be found in Appendix A.

Games such as *Monopoly* do not have an absorbing state but can also be analyzed using Markov processes. A Markov process is regular if some power of its transition matrix contains only positive entries. In these cases, the result used is the Markov Stability Property.

Markov Stability Property: Given a regular Markov process with a transition matrix $\mathbf{P}$, then

$$\lim_{k \to \infty} \mathbf{P}^k = \mathbf{W}$$

where all columns of $\mathbf{W}$ are identical and all entries are positive. Given any distribution of starting locations in a column matrix $\mathbf{u}$,

$$\lim_{k \to \infty} \mathbf{P}^k \mathbf{u} = \mathbf{w}$$

where $\mathbf{w}$ is the common column of $\mathbf{W}$.

CHAPTER 6

Game Theory

Figure 6.1: Components from *Nemesis*.[1]

In *Nemesis*, players take the role of characters attempting to escape a damaged spaceship while being stalked by alien intruders, as shown in Figure 6.1. In this game, players must work together to keep the ship from being destroyed, as this event will cause all players to lose. However, each player is also attempting to complete a personal objective. Assume that the captain and pilot remain on the ship and that the captain will make the first move and can either act selfishly or work to benefit both players.

Is there any way the pilot can entice the captain to work to benefit both players?

 DOI: 10.1201/9781003383529-6

INTRODUCTION

It may be surprising that it has taken 116 pages in a book about games to reach a discussion of game theory. The mathematical subject of game theory focuses on a "rational" player's choices when given a well-defined set of options. This type of analysis is very effective in analyzing games and mechanisms characterized by sequential movement and perfect knowledge (typically referred to as combinatorial games) such as *Chess* and *Go*.

This chapter opens with a discussion of rationality along with the type of results that game theory produces when discussing rational players playing a game. Most tabletop games under discussion in this book cannot be thoroughly analyzed in this way due to their complexity. However, it can still be used to analyze small-scale player interaction within a larger context.

In particular, an understanding of threats and promises can help describe the processes in many games. Two games are heavily associated with threats and promises in the literature (both in game theory and psychology). These are the games of *Chicken*, where threats are effective, but promises are not, and *Prisoner's Dilemma*, where promises are effective and threats are not.

6.1 RATIONALITY

Throughout this chapter, references to game theory will mean the academic study of game theory often taught in mathematics, economics, or philosophy departments. In this field of study, the term rational is frequently used in a specialized sense, and this book will follow this convention. Here, a collection of preferences is said to be rational if it is transitive and any two outcomes can be compared. A rational player (sometimes referred to as a rational agent) is a player who has rational preferences and assumes that their opponents are also rational. This rational player is often referred to as homo economicus (economic man) to distinguish their decisions from those a human might make.

Since homo economicus will assume that their opponent is also homo economicus, they are working under the assumption that their opponent will take advantage of any misstep on their part. This leads to a minimax strategy, where homo economicus will attempt to minimize the maximum possible loss. The same strategy is also referred to as a maximin strategy if phrased as maximizing the minimum possible gain.

There are several reasons why a player may make moves that are not rational. The most obvious reason is that in most popular games, the full consequences of a move cannot be determined. Once a player's opponent is not playing optimally, there can be good reasons to adjust one's strategy to benefit from the opponent's misplay.

Another reason is the result of human psychology. While a minimax strategy is used for some decisions, other strategies are employed in other situations. For example, a player may take great joy in pulling off a high-risk, high-reward move because of its excitement, even if a more conservative move might be more likely to win. Or a player may opt for a more conservative move in places with a significant potential downside. Delving into these questions moves from game theory into human psychology, an interesting subject on its own but not the focus of this book. Appendix A contains references to materials that explore this topic.

Game theorists will talk about a game being solved. A game is strongly solved if one can determine the optimal moves of each player for any game state. For example, the game of *Tic Tac Toe* is strongly solved as, given any game state, most experienced players can readily determine a move that will lead to the best possible outcome for them from the current state. It shouldn't be surprising that strongly solved games soon lose their appeal among experienced players.

A game is weakly solved if one can determine the optimal moves of each player from the starting game state. English Draughts, a form of *Checkers* played on an 8 × 8 board, is an example of a weakly solved game. Even if both players know the weak solution of a game, some of these games still provide exciting gameplay. This can be done by creating a starting state that does not match the standard starting state and having players play the game from that state. If the game provides a handicapping system, players can bid for the role of starting player in these non-standard states.

Finally, a game is ultra-weakly solved if one can determine the outcome of optimal play from the starting game state, but the optimal moves themselves are not known. One weakly-solved game is the game of *Hex*, where it can be proven that the first player will always win in optimal play. The game *Hex* is similar to *Tic Tac Toe*, but it is played on a hexagonal board like the one shown in Figure 6.2a. Like *Tic Tac Toe*, players alternate turns to place pieces on the board. The black player will win if they can connect one of the three spaces on the northwest side of the board with one of the three spaces on the southeast side of the board. The white player will win if they can connect one of the three spaces on the northeast side of the board with one of the three spaces on the southwest side. In the game shown in Figure 6.2b, black played first and white second. It is the white's turn, and no matter where they place their piece, black will be able to win on their next move. Despite it being known that the first player should always win in perfect play, winning strategies are only known for relatively small board sizes.

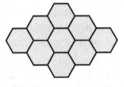

(a) Starting Board.

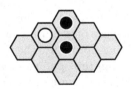

(b) A Winning Position for Black.

Figure 6.2: A Game of *Hex*.

The issue with solving these games is that the decision trees required to analyze most games (presented in Section 6.2) grow exponentially with the number of moves played. The task is manageable for short games with a limited number of moves (for example, there cannot be more than nine moves in a game of *Tic Tac Toe*). However, the average number of moves in a well-played 11×11 game of *Hex* or a game of *Chess* is roughly 40, and in a well-played *Go* game is roughly 200. This high number of moves and the exponential growth of potential states make a brute-force analysis of these games impossible.

6.2   SEQUENTIAL MOVE GAMES

A game is a sequential move game if each player completes their turn before the next player begins their turn. In these games, players must often analyze hypothetical situations to determine the outcome of their chosen action. This is possible under the assumption that both players are homo economicus and that all information is available to both players. The most common tool in handling this situation is the extensive form of the game.

The extensive form of a game presents the choices using a rooted tree where each vertex is a game state, and each edge represents a choice by a player. An edge goes from one vertex to another if a player's decision can cause the game to transition from the state at the edge's tail to the state at the edge's tip.

In the game *Azul*, players take turns selecting colored tiles from piles in the center of the table by selecting a color of tile and then taking all tiles of that color available in that pile. Tiles are then played on each player's board to score victory points. Let's assume that in a two-player game, there is a single pile with two blue tiles, three red tiles, and four teal tiles in the center of the board. Since there are three colors in the center of the board, the first player will end up with two colors, and the second player will end up with one. The goal in *Azul* is to score more points than one's opponent, so the first player will attempt to maximize the difference between their and their opponent's scores. Assume that the difference between player's scores is shown in Figure 6.3.

From this, the best outcome for the first player will be ending up with the red and teal tiles, leaving the opponent with the blue tiles. Can the first player ensure they end up with both colors? Should they select the red or the teal tiles on their move?

First Player Takes	Difference in Points	
blue, red	+3	(benefits first player)
blue, teal	−2	(benefits second player)
red, teal	+4	(benefits first player)

Figure 6.3: Example from *Azul*.

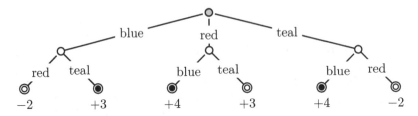

Figure 6.4: Extensive Form for Moves in *Azul*.

The extensive form of this situation is shown in Figure 6.4. The root vertex is at the top of the tree (and shaded gray), and the first player will choose among the three branches emanating from this vertex (selecting either the blue, red, or teal tiles). This transitions the game state to one of the three vertices on the second row. At this point, the second player can choose one of the two remaining colors. This will transition to the third row. At this point, the first player will receive the remaining tiles. This final decision is not included in the tree to save space, but there is only one choice. A counting argument confirms that all possibilities are represented (there are $3! = 6$ ways to order the colors, matching the six final vertices).

The final vertices are indicated with double borders and labeled with the point differences. A positive difference benefits the first player, and a negative one benefits the second. For example, if the first player chooses the blue tiles, followed by the second player choosing the red tiles, the difference will be -2, but if the first player chooses the blue tiles, followed by the second player choosing the teal tiles, the difference will be $+3$.

If the first player chooses the blue tiles, the second prefers the red tiles (obtaining a more negative difference). Since the second player is homo economicus, they will never choose tiles with a more positive value, so the second player will never choose to select the teal tiles if the first player chooses blue tiles. This is indicated by shading the vertex associated with selecting the teal tiles black since that vertex will not be chosen in rational play. A vertex that a player would never choose on their turn is referred to as inactive. Vertices that a player might choose are active strategies. The second player makes the second move (leading to the third row), so the vertices with more positive values on the third row are inactive and shaded black.

The next step is to remove these vertices from the graph, a process known as pruning the tree. The result of the first pruning is shown in Figure 6.5. Having removed all but one choice for the second player, the leaf vertices can be omitted (as when there was only one choice for the first player). Next, the vertices on the second row that are less beneficial for the first player are shaded black to indicate that they will never be selected. This produces the tree shown in Figure 6.6. Repeating this process for the first player's decision will prune the leaves for blue and teal and leave the red leaf with a difference of $+3$, as shown in Figure 6.7.

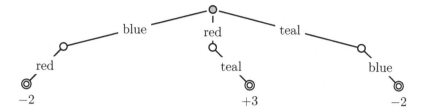

Figure 6.5: Pruning Moves (Part 1) in *Azul*.

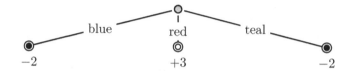

Figure 6.6: Pruning Moves (Part 2) in *Azul*.

Figure 6.7: Pruning Moves (Part 3) in *Azul*.

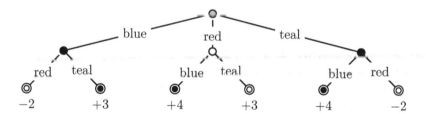

Figure 6.8: Analyzed Extensive Form for Moves in *Azul*.

In actual practice, pruning does not require rewriting the graph but only focusing on the pertinent portions of the graph. The result of this process leaves a diagram like the one in Figure 6.8. The choices made by the players are indicated by following the active nodes (those not shaded black) from the top of the tree to the outcome. Here, the tree indicates that the first player should take the red tiles, the second player should take the teal tiles, and the first player will take the blue tiles. This is the best that the first player can hope for in perfect play.

Decision trees like this work well for short games with only a few available choices or to analyze the final few turns of a longer game. However, even when they cannot be implemented due to their complexity, these trees are an effective way to use graph theory to discuss strategy decisions.

Decision Trees Against Nature

Pruning a decision tree requires knowing how each player will act. However, in many modern tabletop games, the response to a decision is determined by chance. This can happen when cards are drawn, or dice are rolled to determine the consequences of the actions. Random moves also occur in games when a player cannot determine the rational move and applies some other criteria for selection. Being random does not mean that all moves are equally likely, as experienced players will tend to play better moves more frequently. Games in which an opponent (or the environment) makes random decisions are referred to as games against nature. The discussion of probability in Chapter 5 will be applied to these situations.

Returning to the game of *Azul* shown in Figure 6.4, assume that the second player will choose the color they take randomly based on their mood and color preferences (with no consideration of the victory points). Based on a long history of plays against them, the first player believes that the second player will choose teal over either blue or red eight out of ten times, and they will choose blue over red nine out of ten times. This information is added to the decision tree by replacing the color choice with these probabilities, as shown in Figure 6.9. Notice that none of the final states are shaded black because the second player may select to transition the game state to any vertex in the third row.

In this case, homo economicus will weigh the value of the outcomes based on these probabilities. So if they select red (the rational choice if the opponent were playing rationally), there is a 20% chance that the outcome will be +4 and an 80% chance that the outcome will be +3. Since these values are ratio data (as four points are twice as good as two points), it is reasonable to average them. So the expected value of selecting red is a $(0.2)(+4) + (0.8)(+3) = 3.2$ difference. Of course, selecting red cannot result in a difference of 3.2 points

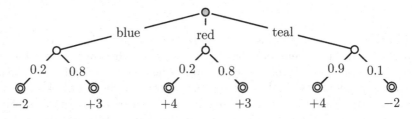

Figure 6.9: Extensive Form for Random Moves in *Azul*.

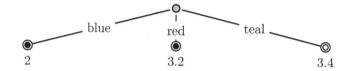

Figure 6.10: Analyzed Extensive Form for Moves in *Azul*.

in a single play. Still, over the long run, a consistent strategy of selecting red will result in an average gain of 32 points over ten instances. The tree shown in Figure 6.10 results from repeating this analysis for the remaining two choices. Pruning the two poor choices, teal has the highest expected value of 3.4 victory points. The net loss for the second player as a result of their irrational play is 0.4 victory points (since optimal play still results in a score difference of +3).

There may be reasons to make a less-than-optimal choice over the short term to benefit in the long term. Even in these cases, estimating the difference between the points earned in a non-optimal move when compared to an optimal move can inform these decisions. In this case, the second player may believe that the teal tiles set them up nicely for the next round, where they hope to earn back the 0.4 points they have surrendered in this round.

6.3 SIMULTANEOUS MOVE GAMES

Some games require simultaneous decisions by both players. This can happen when both players select their choice in secret and simultaneously reveal their choices. One example of such a situation is combat in the game of *Scythe*. In this game, players play different countries in an alternate history of the 1920s. While several activities are available to the player to win the game, one aspect of the game involves using mechanized war machines to gain control over regions of the map. Here, opposing players simultaneously (and secretly) select how much of their current power they wish to spend in the combat, and the player spending the most power will win (ties go to the attacker). Both players lose the power they exert (so this is an example of an all-pay auction, to be discussed in Section 7.3). While the primary goal is to win the battle, players have a secondary goal of retaining as much power as possible for future battles. As a final consideration, if the losing player exerts at least one power, they can draw some combat cards (providing them an advantage in future combats). Suppose that Rose and Colin enter a battle where Rose has a power of two, and Colin has a power of three. Assume both players prefer to earn a combat card if they lose the battle. How much power should each player exert?

A technique similar to pruning trees can remove non-optimal strategies in games, where players make simultaneous moves. However, when each player must make their move without knowing their opponent's move, an alternative

Rose	Colin (defender)			
(attacker)	0	1	2	3
0	R	C	C	C
1	R	R	C	C
2	R	R	R	C

Figure 6.11: Normal Form for a Combat in *Scythe*.

Rose	Colin Plays			
Plays	0	1	2	3
0	R →	C ←	C ←	C
	↑	↓	↓	↓
1	R →	R →	C ←	C
	↑	↑	↓	↑
2	R →	R ←	R →	C

Figure 6.12: Annotated Normal Form of a Combat in *Scythe*.

presentation, the normal form, is used. Again, the focus here will be on two-player games. In this case, the outcomes are placed in a table, where one player's decisions determine which row of the table is used, and the other player's decisions determine which table column is used. In this case, there is no first or second player, and the player who decides the row will be referred to as "Rose" and the player who decides the column as "Colin."

The normal form of the combat scenario described above is shown in Figure 6.11, where "R" indicates a Rose victory and "C" indicates a Colin victory.

The next step is to identify strategies that each player would never utilize. One technique to make this easier is to place arrows between the adjacent outcomes to indicate a preference. Since Rose determines which row will be played, vertical arrows are placed based on Rose's preferences (with the arrow pointing toward Rose's preferred outcome). Similarly, Colin determines which column will be played, so horizontal arrows are placed based on Colin's preferences (with the arrow pointing toward Colin's preferred outcome). This is shown in Figure 6.12.

Notice that the arrows between the column associated with Colin spending zero power and Colin spending one power all point toward spending one power (shown in gray in Figure 6.12). This indicates that Colin will never opt to spend zero power (a guaranteed loss with no combat card). When one strategy (here, spending one power) is always better than a second strategy (here, spending zero power), regardless of what the opponent decides to do, the better strategy dominates the poorer strategy. Here, Colin's strategy of spending one power dominates the strategy of playing zero power. A strategy weakly dominates another strategy if it is never worse than the other strategy

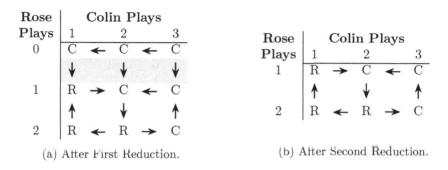

Figure 6.13: Reduced Normal Form for a Combat in *Scythe*.

Rose Plays	Colin Plays 2	3
1	C ←	C
	↓	↑
2	R →	C

Figure 6.14: Completely Reduced Normal Form for a Combat in *Scythe*.

but, in some cases, is better than the other strategy. When one strategy dominates another strategy, the dominated strategy will never be played by homo economicus and can be removed from the table. As a result, the table can be reduced by removing the column for spending zero since Colin will never play it. The reduced table is shown in Figure 6.13a. Now that Colin's strategy of spending zero is ruled out, Rose can also rule out her strategy of spending zero because her strategy of spending one dominates it. Figure 6.13b shows the result of removing Rose's strategy of spending zero.

At this point, no strategy is dominated by an adjacent strategy. However, while comparing only adjacent strategies was convenient, one must compare each strategy with every other strategy. Comparing Colin's strategy of spending one against Colin's strategy of spending three, Colin will always prefer to spend three because it guarantees a victory. In contrast, the strategy of spending one guarantees a defeat. This reduces the table to the table shown in Figure 6.14.

At this point, the table is fully reduced. The remaining active strategies are Colin spending two or three and Rose spending one or two. To make a winning play, one must anticipate their opponent's decision.

Mixed Strategies

In *Rising Sun*, each player controls a clan in legendary feudal Japan. A significant component of this is battling over provinces to win victory points for

Rose	Colin Plays			
Plays	Seppuku	Hostage	Ronin	Poets
Seppuku	2	2	2	−1
Hostage	−3	2	0	−1
Ronin	3	0	1	−3
Poets	0	0	1	2

Figure 6.15: Normal Form for Moves in *Rising Sun*.

Rose	Colin Plays	
Plays	Hostage	Poets
Seppuku	2	−1
Poets	0	2

Figure 6.16: Reduced Normal Form for Moves in *Rising Sun*.

control of the province. In the game, an all-pay auction (see Section 7.3) takes place before battle as players determine which strategies they intend to pursue. Rose and Colin have four strategies that they can pursue during the auction. These are committing seppuku (killing all of the player's units in that region but receiving victory points for each death), taking a hostage (capturing one of the opponent's units), recruiting ronin (recruiting extra forces), and hiring imperial poets (earning victory points for each unit killed by commemorating them in poems). While Rose and Colin can spread their efforts across all four strategies in the game, assume that each player will put all of their efforts into a single strategy. Assume that the difference between the number of victory points Rose and Colin will earn is presented in Figure 6.15. Positive values indicate a benefit to Rose, while negative values indicate a benefit to Colin. What is the optimal strategy for Rose in this situation? What is the optimal strategy for Colin in this situation?

Reducing this table by removing dominated strategies leads to the table displayed in Figure 6.16. In games where the outcomes consist of interval or ratio data, players can employ a mixed strategy to determine their move. As odd as it sounds, analyzing this situation from Rose's perspective is the best way to determine Colin's optimal play. Rose does not know the percentage of time Colin will take a hostage and the percentage of time Colin will hire poets, so she assigns a probability of p to Colin taking a hostage and $1 - p$ to Colin hiring poets. If Rose chooses to commit seppuku, her payoff will be $(p)(2) + (1-p)(-1) = 3p - 1$. On the other hand, if Rose chooses to hire poets, her payoff will be $(p)(0) + (1 - p)(2) = 2 - 2p$. These two lines are plotted in Figure 6.17, note that they intersect at the point $(3/5, 4/5)$.

If the value of p is less than $3/5$, Rose should choose to hire poets as this will result in a higher expected value than committing seppuku. On the other hand, if the value of p is more than $3/5$, Rose should commit seppuku as this

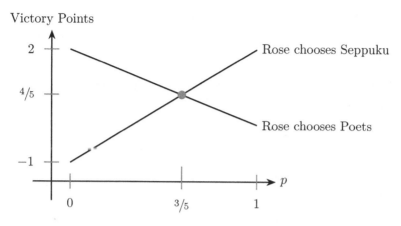

Figure 6.17: Payoff to Rose if Colin Takes a Hostage with Probability p in *Rising Sun*.

will result in a higher expected value than hiring poets. If Colin sets p at $3/5$, he guarantees that optimal play by Rose will earn the lowest expected payoff, $4/5$ victory points. From Colin's perspective, the decision to play $p = 3/5$ has minimized the maximum payoff to Rose. Completely deterministic strategies (for example, taking a hostage every time) are pure strategies. Strategies that blend pure strategies using some form of randomness are mixed strategies. So Colin's optimal strategy is a mixed strategy that takes a hostage $3/5$ of the time and hires poets the remaining $2/5$ of the time.

From Rose's perspective, since both of her choices lead to the same payoff, any mixed strategy using them will also lead to the same payoff. In effect, Colin has chosen a strategy that eliminates Rose's ability to affect the payoff. The goal for Rose is not to maximize her payoff (since Colin's strategy sets the payoff at $4/5$) but to choose a mixed strategy that Colin cannot exploit. Repeating the same process from Colin's perspective leads Rose to hire poets with a probability of $q = 3/5$, as shown in Figure 6.18. The symmetry in the answer (that $p = q$) arises due to some symmetry in this particular scenario, which may not exist in every scenario.

Plotting the expected victory points as a function of Colin's probability of taking a hostage, p, and Rose's probability of hiring poets, q, leads to the graphs shown in Figure 6.19. The graph in Figure 6.19a is a heat map with darker shades in lower regions (more favorable to Colin). The lowest point occurs at $(p, q) = (0, 0)$ and corresponds to the payoff of -1. The two high points at $(p, q) = (0, 1)$ and $(p, q) = (1, 0)$ correspond to the two outcomes with a payoff of $+2$. The two dark gray lines are the contours, where the payoff is equal to $4/5$, and their intersection is the result of rational play. The graph shown in Figure 6.19b is the shape when viewed from the point $(p, q) = (0, 0)$ looking toward the equilibrium (along the arrow shown in the heap map). The

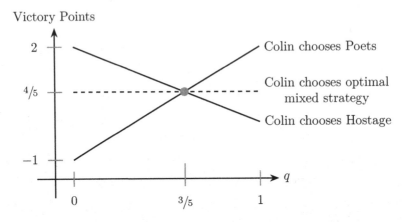

Figure 6.18: Payoff to Colin if Rose Hires Poets with Probability q in *Rising Sun*.

(a) As a Heat Map.   (b) As a Saddle.

Figure 6.19: Payoffs in *Rising Sun*.

shape of the graph resembles a saddle, and the optimal play $(3/5, 3/5)$ is often referred to as a saddle point.

Suppose Colin decides to adjust his probability away from $p = 3/5$, moving left or right in the region in Figure 6.19a. In that case, Rose will be able to respond by shifting forward or backward from $q = 3/5$ in a way that moves toward one of the two higher corners, increasing her payoff. Similarly, suppose Rose decides to adjust her probability away from $q = 3/5$, moving forward or backward on the graph. In that case, Colin will be able to respond by shifting left or right from $p = 3/5$ in a way that moves toward one of the two lower corners, decreasing the payoff to Rose.

A Nash equilibrium[2] is an outcome from which no player has any incentive to deviate unilaterally. If the outcome involves a non-deterministic component outside of the control of one's opponent, the optimal strategy is sometimes

---

[2]The Nash equilibrium is named after the twentieth-century American mathematician John Nash, who received the Nobel Memorial Prize in Economics for his work in game theory.

called a Bayesian-Nash equilibrium[3] (see page 143 for an example). In a mixed strategy, the Nash Equilibrium has players playing each strategy with a probability that equalizes all of the expected payoffs. In effect, each player chooses a strategy that removes the ability of their opponent to affect the payoff. This is a weaker solution than finding a single strategy that dominates all other strategies. In the case of a dominating strategy, a player should always play the dominating strategy regardless of what their opponent plays. In a Nash equilibrium, a player should play to the equilibrium *if* they believe their opponent is also playing optimally (and so will also be playing to the equilibrium). Before moving on to more complicated games, it is worth noting that a Nash equilibrium will always exist in a finite normal form game.

Nash Existence Theorem: Any normal form game with a finite number of strategies for each player has a Nash equilibrium, which can be obtained by all players playing mixed strategies.

6.4 THREATS AND PROMISES

While it is often good to make the first move, the player who moves second may be able to persuade the first player to choose a different course of action through threats and promises. Both of these are employed by the player making the second move to induce the player making the first move to re-evaluate the potential outcomes. In particular, the second player proposes that they will not act as homo economicus but instead play irrationally.

All the previous games in this chapter are constant-sum games. These are games where every favorable payoff to one player is balanced by an unfavorable payoff to the other player(s). In these games, the outcome could be reduced to knowing the payoff to one player (as the other payoff will be equal and opposite). More complicated games allow for win-win or lose-lose situations. In these games, payoffs must be listed for all players. For a two-player game, payoffs are typically given as pairs of numbers. In sequential games, the first number represents the payoff to the first player, and the second number represents the payoff to the second player. In simultaneous games, the first number represents the payoff to Rose, and the second number is the payoff to Colin. When payoffs are listed for all players, the convention will be that higher values are better. So a payoff of $(2, 3)$ would benefit Rose more than a payoff of $(1, 4)$ because $2 > 1$. A payoff of $(1, 4)$ would benefit Colin more than a payoff of $(2, 3)$ because $4 > 3$. When looking at these payoffs, it is important that one only consider the value associated with the player making a choice. For example, when deciding between an outcome with a payoff of $(1, 2)$ and an outcome with a payoff of $(2, 4)$, Rose only considers the fact that $2 > 1$,

---

[3]The Bayesian-Nash equilibrium is named after John Nash and Thomas Bayes. Thomas Bayes was an eighteenth-century English statistician who developed Bayes' Theorem, which can be used to compute conditional probabilities.

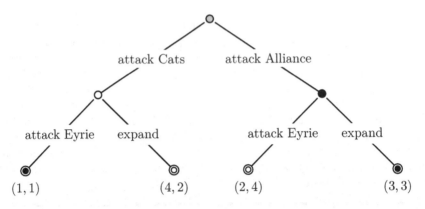

Figure 6.20: Extensive Form of a Turn in *Root*: Eyrie to Make First Move, Cats to Respond. Payoffs are Listed as (Payoff to Eyrie, Payoff to Cats).

ignoring the fact that Colin appears to be getting more of an advantage from the second outcome. Any benefit to Colin was already considered when the values were assigned as Rose's payoff. Unless otherwise noted, these payoffs should be treated as ordinal data, so taking averages is not meaningful.

Threats

In the game of *Root*, players take the roles of woodland animal factions in a struggle for dominance over the wilderness. Despite its seemingly light theme, the game is a complicated asymmetric area control game. The Eyrie Dynasties, the Marquise de Cats, and the Woodland Alliance are three factions in the game, and the Eyrie must decide whether to launch an attack on the Cats or the Alliance. After the Eyrie makes their move, the Cats must decide whether to attack the Eyrie or expand into an uncontrolled forest region. The decision tree is shown in Figure 6.20. An example where these preferences would arise is where the Eyrie and the Cats are two powerful, evenly-matched factions while the Alliance is a less powerful faction. If the Eyrie and Cats enter an all-out war against each other, the Alliance will likely be able to leap-frog both players in the resulting vacuum. However, the Eyrie and Cats would each benefit if they attacked the other faction without being counterattacked. The figure indicates these rankings as pairs of numbers, with the payoff to the Eyrie in the first place and the payoff to the Cats in the second place. In this case, each faction would like their payoff to be as large as possible. If both players are rational, then the Eyrie will choose to attack the Cats, and the Cats will choose to expand to an uncontrolled region. This is the Nash equilibrium in the game, as neither player would choose to deviate from these strategies unilaterally.

However, the Marquise de Cat can threaten: "If you attack me, I will attack you back" as indicated by the dashed threat arrow in Figure 6.21. Assuming

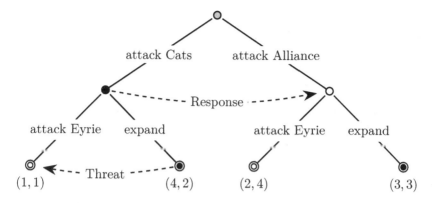

Figure 6.21: Cats Make a Threat in *Root*.

the Eyrie believes the Cats, the Eyrie is no longer comparing the $(4, 2)$ outcome against the $(2, 4)$ outcome but is comparing the $(1, 1)$ outcome against the $(2, 4)$ outcome. In this situation, the rational choice for the Eyrie will be to attack the Alliance instead. The Cats' deviation from the Nash equilibrium is predicated on the fact that it will also cause the Eyrie to deviate.

What makes this a threat in game theoretic terms is that the threatening player (the Cats) is claiming that their response to the rational choice of the threatened player (the Eyrie) will hurt both players' outcomes. This threat is effective because if it is believed, then it will result in a rational player changing their choice (in this case, the Eyrie choosing to attack the Alliance instead of the Cats). Finally, this threat is beneficial (to the threatening player) because after the Eyrie changes their decision, the Cats will end up in a better situation than the outcome of rational play.

How the Cats convince the Eyrie that they will carry out their threat is largely non-mathematical. The Cats must convince the Eyrie that their preferences differ from what is shown in the decision tree. If the Cats convince the Eyrie that they would instead respond to an attack by counter-attacking (perhaps because they don't want to feel like an easy target), the Eyrie will do better to attack the Alliance. As a result, threats are most effective if the threatening player has demonstrated that they do not bluff. By following through on low-stakes threats in the past, a player has set themselves up to be believed on a high-stakes threat in the future. Similarly, by not responding to low-stakes threats, the threatened player can establish that threats will be ineffective against them.

Before moving to promises, notice that a threat is a fundamentally aggressive play. The threatened player will have a less desirable outcome than the rational outcome no matter their choice (assuming the threat will be carried out).

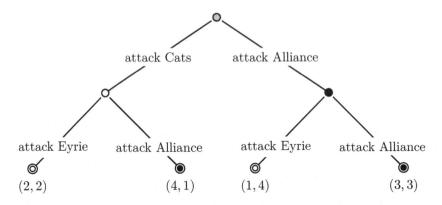

Figure 6.22: Extensive Form of a Turn in *Root*: Eyrie to Make First Move, Cats to Respond. Payoffs are Listed as (Payoff to Eyrie, Payoff to Cats).

Promises

A promise is a counterpart to a threat. Rather than responding to a rational choice of the first player, a promise tries to entice the first player to make an irrational decision in the hopes of a better pay-off. Unlike a threat, no choice by the first player will result in a less desirable outcome than the rational equilibrium.

Returning to *Root*, consider the decision tree shown in Figure 6.22. The top two outcomes for each faction are to avoid being attacked. However, in this scenario, unlike the previous scenario, the assumption is that there is little chance that the Alliance will be able to leap-frog either player even if they attack each other. As a result, the Eyrie and Cats would prefer to attack each other over attacking the Alliance. Again, this is the Nash equilibrium

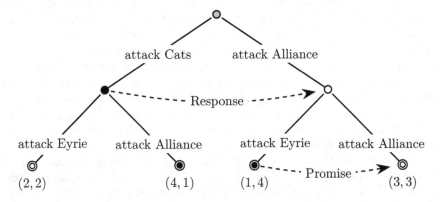

Figure 6.23: Cats Make a Promise in *Root*.

	Swerve	Straight
Swerve	(3, 3)	(2, 4)
Straight	(4, 2)	(1, 1)

(a) *Chicken.*

	Silent	Talk
Silent	(3, 3)	(1, 4)
Talk	(4, 1)	(2, 2)

(b) *Prisoner's Dilemma.*

Figure 6.24: Games of *Chicken* and *Prisoner's Dilemma.*

in the game, as neither player would choose to deviate from these strategies unilaterallly.

In this situation, the rational equilibrium would be for the Eyrie to attack the Cats, followed by the Cats attacking the Eyrie. The Cats cannot improve this with a threat (such as "If you attack me, I will attack the Alliance" because this would lead to an even better outcome for the Eyrie). However, there is an outcome that benefits both the Eyrie and the Cats more than the rational equilibrium. Seeing this, the Cats may respond: "If you attack the Alliance, I will also attack the Alliance." The results of this promise are shown in Figure 6.23. If the Eyrie trusts that the Cats will not renege on their promise, they will attack the Alliance. This is a promise because the promising player is offering not to play rationally if the first player chooses not to play rationally. The promise is effective because this will result in the first player choosing a different option. Finally, the promise is beneficial because the promised outcome is more desirable to the promising player than the rational equilibrium. And, like threats, whether the first player believes the promising player depends on past play between these players.

Finally, returning to the situation in Figure 6.21, notice that while a promise is possible (such as "If you attack the Alliance, I promise to expand instead of attacking you"), it will not be effective because the Eyrie will receive their top choice if they attack the Cats and the Cats cannot promise more than that.

6.5 TWO WELL-STUDIED GAMES

The examples of threats and promises in Section 6.4 are related to two games extensively studied in game theory: *Chicken* and *Prisoner's Dilemma.* While these games are traditionally presented as simultaneous move games, Figure 6.21 matches the game of *Chicken*, while Figure 6.23 matches the game of *Prisoner's Dilemma.* The normal form of each of these games is shown in Figure 6.24.

When analyzing games such as these, it is helpful to use two concepts associated with Vilfredo Pareto.[4] One outcome is a Pareto improvement if

---

[4]Vilfredo Pareto was a nineteenth–twentieth-century Italian economist whose economic analysis found applications in game theory.

it is not worse for any players while benefiting some players. An outcome is Pareto optimal if it cannot be Pareto improved.

The Game of *Chicken*

The typical story associated with *Chicken* involves two dueling teenagers who drive toward each other in their cars. Each teenager may swerve (swerving their car to the side and out of the way of the other car) or continue straight. If neither car swerves, the cars will crash, causing severe injury to both teenagers. On the other hand, if one teenager swerves while the other continues straight, the teenager who has continued straight wins the duel. Finally, the duel is considered a tie if both cars swerve.

In this game, the Nash equilibrium occurs when the players choose different strategies, and every outcome other than both players going straight is Pareto optimal. Every time rational players play the game, it should result in one of the two equilibria from which there is no Pareto improvement. In the simultaneous play of *Chicken*, threats are effective and beneficial as a threat can remove one of the Nash equilibria from play. If a player can convince their opponent before the start of the duel that they will not swerve, then the opponent's rational response will be to swerve. As noted earlier, this threat is aggressive, and the player who makes a convincing threat first can go straight while their opponent swerves.

The Game of *Prisoner's Dilemma*

Prisoner's Dilemma is typically presented as a story about two prisoners accused of participating in a crime together. It assumes that there is already adequate evidence to convict them of a crime (for example, robbery) but not enough evidence to convict them of a more serious crime (for example, murder). The prisoners are separated from each other, and each prisoner is presented with the following offer: they can be granted immunity for the robbery if they are willing to implicate the other prisoner in the murder. If both prisoners stay silent, they will both be convicted of robbery. If only one prisoner chooses to talk to the police, they will be granted immunity, and the other prisoner will be convicted of murder. If both criminals implicate each other, they can expect an intermediate conviction of manslaughter as they have provided the police with a stronger case against both, but each can claim that the other was the primary culprit. For each prisoner, the best outcome would be immunity, followed by a conviction for robbery, then manslaughter, then murder. In *Prisoner's Dilemma*, "cooperating" refers to cooperating with the other prisoner (staying silent), while "defecting" refers to implicating the other prisoner. Suppose one prisoner defects while the other prisoner cooperates. In that case, the prisoner who cooperates (and is convicted of murder) is often referred to as the sucker, as they have trusted their partner who has turned on them.

In this game, the Nash equilibrium is the outcome where both players talk. Interestingly, this is the only outcome that is not Pareto optimal. This means that rational play in this game leads to a non-optimal outcome. In simultaneous play of *Prisoner's Dilemma*, promises are effective and beneficial. If both players promise to stay silent (and trust each other), they can obtain a better outcome than if they play rationally. As mentioned earlier, promises are a friendly agreement, so, unlike threats, there is no need to be the first to make a promise as long as both players make the promise before playing the game.

Semi-cooperative Games

In a cooperative game, all players work together toward a common goal and will win or lose together. In a semi-cooperative game, all players can still lose if a common goal is not achieved, but each player will also have personal objectives that must be completed as an additional condition to win. The dynamics of semi-cooperative games often mimic that of *Chicken* or *Prisoner's Dilemma*.

Nemesis (p.116) is a semi-cooperative game. Although a player's objective may not be directly confrontational, players will always be torn between choosing an action that prevents the destruction of the ship and progressing toward their personal objective. For example, the captain and the pilot may be in separate rooms where they can each spend an action to repair the ship (aiding both players) or to complete a personal objective (aiding only themselves). Assume that if neither character repairs the ship, there is a high probability that both players will lose, as the ship will be destroyed before either character can escape. However, suppose only one character, the captain, repairs the ship. In that case, the ship will remain intact long enough for the pilot to escape but not long enough for the captain to complete their task and escape, with the pilot winning and the captain losing. Similarly, if the pilot repairs the ship and the captain completes their objective, the pilot will lose and the captain will win. Finally, if both players repair the ship, it will remain intact for much longer, probably giving both characters time to complete their objectives and escape (and allowing both players to win).

The best outcome for a player would be for their character to pursue their personal objective while the other character repairs the ship. The second best would be for both characters to repair the ship, providing more time to win. The remaining outcomes are more difficult to compare. One of these outcomes would make the player the sucker (choosing to repair the ship while the other character completes their objective). The other outcome would be neither player repairing the ship, in which case they would both lose. Exactly how these are ranked depends on the player's preferences in gameplay. If a player would prefer that no one win if they cannot, then they may prefer to have everyone lose over becoming the sucker and view this game as *Prisoner's*

Dilemma. On the other hand, if a player would prefer that someone survive over everyone dying, they view this game as *Chicken.*

In the opening question of the chapter, if the pilot believes this will play out as *Prisoner's Dilemma*, they will promise: "If you repair the ship, I will also repair the ship." If the pilot believes this will play out as *Chicken*, they will threaten: "If you don't repair the ship, neither will I." More likely, the pilot may be unable to determine the captain's preferences and provide both a threat and a promise: "If you repair the ship, so will I, but if you don't repair the ship, neither will I." If the captain believes the pilot is not bluffing, it is in the captain's best interest to work to benefit both players.

The strategy of repeating the opponent's play is often referred to as a "tit for tat" strategy. There is significant research into how humans respond to this strategy when placed into repeated plays of *Prisoner's Dilemma*, see Appendix A.

CHAPTER 7

Auctions

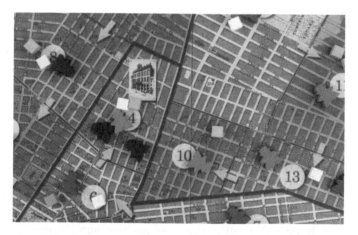

Figure 7.1: Components from *Tammany Hall*.[1]

Games that simulate elections or legislative processes often allow players to spend influence to affect the outcome. In *Tammany Hall*, players can bid influence to win elections in the wards where they have built up their network, shown in Figure 7.1. All influence spent is lost whether the player wins the election or not.

In a five-player game of *Tammany Hall*, an election is held whose value to the players is between zero and ten influence. How much should be bid by a player who values a victory in the election as being worth eight influence?

---

[1]Tammany Hall images and text were used with permission from StrataMax Games and Pandasaurus Games. They are copyright Strata Games and/or Pandasaurus Games, all rights reserved worldwide.

DOI: 10.1201/9781003383529-7

INTRODUCTION

Auctions occur in a variety of forms in tabletop games. Some games, like *Modern Art*, present traditional auctions, where the price rises until only one bidder remains. Other games, such as *Through the Ages: A New Story of Civilization*, present auctions, where the cost of the item decreases until someone purchases the item. The third form of auction discussed are those found in games like *Tammany Hall*, where even the losing players must pay their bid's value.

Economists have studied auctions extensively, as auctions are used in many situations with significant real-world impacts. However, some differences exist between the auctions studied by economists and those that appear in tabletop games. One difference is that an economic auction aims to efficiently establish the appropriate market value of an item for which there are few or no comparable items. However, the game designer frequently desires to provide the player with difficult decisions, which often means an inefficient auction system where traditional strategies may not be optimal. Another difference is that the auctioneer aims to sell the item with the highest revenue in an economic auction. In tabletop games, the auctioneer may be indifferent to the profit to be made on the item.

7.1 ENGLISH AND VICKREY AUCTIONS

Several assumptions will be made throughout this chapter to make the analysis easier to follow, but these can be removed with more careful work. The first assumption is that all bidders are rational (in the sense of homo economicus) and will all engage in the same strategy. It is assumed that the optimal bidding strategy is continuous, meaning that any positive number can be bid, and monotonic, meaning that higher valuations of an object do not lead to lower bids for that object. Finally, it is assumed that there are no ties in the valuation of the object being auctioned.

English Auction

In *Modern Art*, players take the role of art museums bidding on artwork using different auction types during the game. One of the game's auction types is called an open auction. In this auction, the players place bids in any order, and each bid must exceed the previous bid. When no one is willing to bid more than the current bid, the artwork is sold to the highest bidder, who pays the highest bid price. Assume there are five players, and one player believes the artwork will be worth €20k to them. How high should that player bid?

This type of auction is an English auction, also called an open ascending auction. It is open because bidders can see other bids, and it is ascending because the bid increases as the auction progresses. In this auction, the player must decide when to stop bidding, and it will be shown that the optimal strategy is straightforward: they should continue to bid until they win or until the bidding reaches the value they assigned to the artwork.

The variable θ will represent the value the player assigns to the artwork, the €20k above. One can think of this as the most money the player would pay outside of an auction environment to purchase the item. Let M be the maximum bid for the object among all other players and b be the player's maximum bid. If $b > M$, the player will win the auction when the bidding reaches M, and all other bidders drop out. At this point, the player will pay M and win the item. In this case, the player's gain will be $\theta - M$. If $b < M$, the player will lose to the bidder who values the artwork at M. In this case, the player's gain will be zero. The general goal in auction theory is to use the information about the player's valuation of the item and the number of bidders to determine an optimal value for the player's bid. Here, it will be shown that in an English auction, having a maximum bid of θ is optimal. This will be done by comparing the maximum bid of θ to alternative strategies, where the maximum bid is less than θ and the maximum bid is greater than θ.

In the case that the player's maximum bid, b, is less than θ, there are three cases to consider: $M < b$, $b < M < \theta$ and $\theta < M$. In the first case, the auction ends when the player remains the last bidder, and the item is purchased for the cost of M, so there is no difference between a maximum possible bid of b and one of θ. Similarly, in the third case, whether the player's maximum bid is b or θ, the player will lose the auction with zero gain. So, the only case where the choice of b or θ affects the expected value is if $b < M < \theta$. In this case, a maximum bid of b will lead to a zero gain, as the player will lose the auction. However, a maximum bid of θ will lead to the player winning the auction and paying M for the object. This is a gain of $\theta - M > 0$. Combining all three cases,

$$E[\text{Gain} \,|\, \text{bid of } \theta] - E[\text{Gain} \,|\, \text{bid of } b] = (\theta - M)P[b < M < \theta] > 0.$$

This matches the intuition that if a player bids $b < \theta$, they lose some expected value when they are outbid with a bid $M < \theta$.

The consideration of $\theta < b$ is similar. In the case where $M < \theta$, the player will win the auction and pay M whether their maximum bid is b or θ. On the other hand, if $b < M$, the player will lose the auction whether their maximum bid is b or θ with zero gain. So, the only case that affects the expected value will be when $\theta < M < b$. In this case, the player with a maximum bid of b will win the auction and pay M for the item. However, in this case, $\theta < M$ and so the difference in the expected gain between a maximum bid of θ and a maximum bid of b is $0 - (\theta - M) = (M - \theta) > 0$. Combining these three cases leads to the result that

$$E[\text{Gain} \,|\, \text{bid of } \theta] - E[\text{Gain} \,|\, \text{bid of } b] = (M - \theta)P[\theta < M < b] > 0.$$

Again, the result should match intuition. The player will lose money if they bid $b > \theta$ when $\theta < M < b$ as they have won an auction in which they overpaid for the item.

Returning to the question involving bids in *Modern Art*, the player should make bids up to their valuation. In this case, up to €20k. Generally, a bidder should follow the Optimal Bid in an English Auction.

Optimal Bid in an English Auction: In an English auction, the weakly dominant strategy for a bidder is to have a maximum bid equal to their valuation.

Vickrey Auction

The Vickrey auction[2] is included here because it is a common auction type even though it does not appear in many (any?) tabletop games.[3] In the Vickrey auction, each player makes a single bid. The player with the highest bid will win the auction but will pay the price of the second-highest bid. As a result, the Vickrey auction is often called a sealed-bid second-price auction because the bidders do not see other bids (they are sealed) and second-price because the winning bidder pays the second-highest bid. It can be shown that the Vickrey auction is mechanically equivalent to the English auction. Therefore, each player is incentivized to bid their valuation, knowing that if they win, they will only need to pay the maximum bid of the other players. While mechanically equivalent to an English auction, bidders are less active (which may make it less exciting for tabletop games), and the auction will not reveal the bidders' valuations (as only the auctioneer knows the bids).

7.2 BLIND AND DUTCH AUCTIONS

Blind Auction

In *QE*, players take the role of central banks who are attempting to stimulate their economies through quantitative easing (the Q and E in QE). Players will bid on companies by secretly writing their bids and giving them to the auctioneer. The auctioneer will then determine the highest bid and declare that player the winner without revealing the winning bid to the other players. The interesting twist that *QE* places on this process is that the players have unlimited money, but the player who has spent the most money at the game's end cannot win. Assume there are 4 players, and one player believes the company will be worth \$1M to them. What should that player bid?

The impact of the twist is ignored here to focus on the bidding aspect of the game. A blind auction, sometimes called a sealed first-bid auction, is similar to the Vickrey auction in that everyone submits a single bid, but in

---

[2]The Vickrey Auction is named after William Vickrey, a twentieth-century Canadian-American economist and Nobel laureate who applied game theoretic techniques to analyzing auctions and discovered that a Vickrey auction is equivalent to an English auction.
[3]If you know of a tabletop game that uses a Vickrey auction, please let me know.

this case, the winning bidder pays the value of their bid. Blind auctions are also in *Modern Art* (p.138) as the hidden auction type.

Because this is more complicated than an English auction, some additional assumptions are necessary to determine the solution. The goal is to find a strategy function, s, that converts a personal valuation θ into a bid b. In the case of an English auction, the strategy of $s(\theta) = \theta$ is optimal. Under the assumption that all players are homo economicus, every player will use the same strategy function. What will differ will be their valuations of the object up for auction. All bidder valuations will be assumed to be independent, sometimes referred to as the assumption of independent private values. It is also assumed that all valuations come from a uniform distribution between zero and one to make computations easier. This assumption means if θ is a fixed value between zero and one and X is a random valuation, then

$$P[X < \theta] = \frac{\text{len}([0, \theta])}{\text{len}([0, 1])} = \theta,$$

where $\text{len}([a, b]) = b - a$ is the length of the interval from a to b.

The assumption of independence implies that the probability that $M < \theta$ is equal to

$$P[M < \theta] = P[\theta_1 < \theta] \times P[\theta_2 < \theta] \times P[\theta_3 < \theta] = \theta^3,$$

where θ_1, θ_2, and θ_3 are the valuations of the three other bidders.

The player's goal is to maximize their gain in the auction. Because they make no gain or loss on a lost auction, the expected gain is the product of the probability that the player will win the auction, θ^3, times their gain upon winning, $\theta - s(\theta)$. This expected gain is given in the function y, where $g(\theta) = \theta^3(\theta - s(\theta))$.

For any strategy, a player can adjust their bid up and down by changing strategies or by changing the valuation they pass to the strategy. The monotonicity condition (that higher valuations always lead to higher bids) allows a player with one strategy to mimic any other strategy by manipulating the input to the strategy function. If all players make a symmetrical decision to use a manipulated valuation, v, then the player will win the auction against the other bidders with probability v^3, and the expected gain associated with that valuation is given by the function

$$g(v) = v^3(\theta - s(v)).$$

Notice that this adjusts their probability of winning the auction and the amount they pay upon winning, but not the item's value to them. The valuation v results from the player pretending that the item is worth more or less than its actual value to them.

Since the player may prefer not to manipulate their valuation (and doesn't even know what the correct manipulation would be), the function s should be selected so that g is maximized when the player does not manipulate their valuation and selects $v = \theta$, their accurate valuation. To complete this reasoning, the assumption must be made that s, and therefore g can be differentiated. In this situation, g will be maximized at $v = \theta$ if and only if $g'(\theta) = 0$. Taking a derivative yields a first-order linear differential equation that can be integrated into the solution with the assumption that $s(0) = 0$ used to identify the constant of integration:

$$g'(v) = 3v^2 \times (\theta - s(v)) + v^3 \times (-s'(v))$$
$$0 = 3\theta^2 \times (\theta - s(\theta)) - \theta^3 \times s'(\theta)$$
$$\theta^3 s'(\theta) + 3\theta^2 s(\theta) = 3\theta^3$$
$$\theta^3 s(\theta) = \frac{3}{4}\theta^4$$
$$s(\theta) = \frac{3}{4}\theta.$$

So, in this case, the player should multiply their valuation by a factor of $3/4$ to obtain their bid. In the example from QE, the bid would be $750K. The practice of bidding less than one's valuation in an auction is referred to as bid shading. Handling this in the case of n players gives the general formula for shading a bid in a blind auction shown in Optimal Bid in a Blind Auction.

Optimal Bid in a Blind Auction: In an n-bidder auction in which the valuations are independent and uniform on an interval $[0, 1]$, the Bayesian-Nash equilibrium strategy is to bid

$$s(\theta) = \frac{n-1}{n}\theta.$$

Overall, the connection between the number of bidders and the amount of shading makes sense. In larger auctions, the bidding would be expected to be more competitive, so bidders would need to bid more aggressively to win the auction.

Dutch Auction

In *Through the Ages: A New Story of Civilization*, players attempt to build a flourishing civilization from antiquity through the modern era. Players will collect civil cards to improve their civilization by spending a currency known as civil actions. During a player's turn, there are thirteen civil cards that the player may purchase in a row ordered from left to right. The first (left-most) five cards cost one civil action to purchase, the following four cards cost two civil actions to purchase, and the last four cards cost three civil actions to purchase. In a four-player game, at the start of each player's turn, the player will remove the left-most card, slide all cards left to fill empty spaces, and then draw cards to fill the right-most positions in the row. The player must decide whether a card is valuable enough to take. It may be worth paying two or three civil action points to take a card before another player can.

This mechanism is similar to a Dutch auction[4], also called a descending auction because the bid value decreases as the auction progresses. In a Dutch auction, the price starts above any reasonable bid and slowly decreases until a bidder claims the item at the given price. Like the blind auction, a bidder who bids their valuation will have an expected gain of zero, so bidders must shade their bids to have a positive expected gain. Also, like a blind auction, no information is provided to the bidder about how other bidders have valued the item other than the winning bid. These similarities are not a coincidence, as the Dutch auction is equivalent to a blind auction. So, the optimal strategy for a Dutch auction is identical to the strategy for a blind auction.

Domination and Equilibrium

It is important to note that the English and blind auction strategies respond differently to irrational players. The English auction strategy dominates all other strategies. What the other bidders do is irrelevant: a player should bid up to their valuation. On the other hand, the blind auction strategy is only a Bayesian-Nash Equilibrium. It is Bayesian because it depends on the valuation by the other bidders (which is randomly determined). If the other bidders bid irrationally, then a player should be able to adjust their strategy away from the optimal strategy to take advantage of the other bidders' irrationality.

One can see the difference by considering an auction with two bidders, the player who has a positive valuation $\theta > 0$ and one other bidder who follows the strategy of $s(v) = 0$ (bidding zero in all auctions). In a Vickrey auction (equivalent to an English auction), the player's strategy would not change: they would still bid up to their valuation and receive the item for zero. On the other hand, in a blind auction, it would make little sense for the player to bid anything other than the smallest positive bid allowed since that bid would provide them with the largest gain.

Auctions in Ra

In the game Ra, players seek to expand their power and influence in ancient Egypt by forming collections of tiles, which are earned through an ascending auction. On a player's turn, they may add a random tile to the current lot of tiles or call for an auction, with the auction winner taking control over the current lot of tiles. The auctions are turn-based, with each player making a single open bid in clockwise order and the player who called the auction making the last bid. Bidding is done using thirteen sun tiles, numbered from one to thirteen. All players know which player holds which tiles. In a

---

[4]The Dutch auction is named after the tulip auctions that are held in the Netherlands.

Figure 7.2: Player Board in *Ra*.[5]

three-player game, each player has four sun tiles for bidding, and a winning player will forfeit the tile they used as their bid. Figure 7.2 shows an example of a player board with the sun tiles numbered four, nine, and twelve. The fourth sun tile was used in a previous auction.

Analyzing an auction in this game will demonstrate how game mechanisms cause the auction results to deviate from the economic description. Assume that Rose, Colin, and Larry are playing *Ra* and that Larry has called for an auction in which Rose (holding tiles five and eleven) will bid first, Colin (holding tiles four, nine, and twelve) will bid second, and Larry (holding tiles six and thirteen) will bid third. The bidder valuations are independent and uniformly distributed between zero and 14 (under the assumption that once the valuation has reached 14, some player would have called an auction). The optimal bids deviate from the theory for two reasons: the player bids will occur sequentially rather than simultaneously, and the sun tiles constrain bids. In this subsection, the strategies are guided by the theoretical results and verified by simulating auctions.

At first, players are allowed to bid any value. From Rose's perspective, the auction is a three-player blind auction because Rose won't know Colin's and Larry's bids before making her bid. Rose follows the blind auction bid and shades her bid by $2/3$. Then, from Colin's perspective, this is a two-bidder auction with a lower bid equal to Rose's bid. Colin will not bid if his valuation is below Rose's bid. However, if his valuation is above Rose's bid, he will be in a two-bidder auction with Larry and will shade his bid by $1/2$. This misses the situation where Colin's valuation is above Rose's bid, but his shaded bid would be below Rose's bid. In this case, Colin will bid a small increment above Rose's bid to outbid Rose. Finally, Larry will make the final bid. No one can bid after Larry's bid, so if Larry's valuation is above the current bid, Larry will bid just enough to win the auction.

Evidence that this process is reasonable can be found in the simulation data in Table 7.1. The simulation ran one million auctions with Rose shading her bids by a factor of $i/6$, for i running from one to five, Colin shading his bids by a factor of $j/6$, for j running from one to five. In cases where his valuation

---

Table 7.1: Simulation Results from *Ra*.

Rose's Shading	Colin's Shading				
	$1/6$	$2/6$	$3/6$	$4/6$	$5/6$
$1/6$	$0.13, 0.85$	$0.13, 1.13$	$0.13, 1.21$	$0.13, 1.03$	$0.13, 0.62$
$2/6$	$0.33, 0.94$	$0.33, 1.12$	$0.33, 1.18$	$0.33, 1.01$	$0.33, 0.62$
$3/6$	$0.49, 0.97$	$0.49, 1.08$	$0.49, 1.12$	$0.49, 0.97$	$0.49, 0.60$
$4/6$	$0.54, 0.89$	$0.54, 0.98$	$0.54, 1.02$	$0.54, 0.90$	$0.54, 0.56$
$5/6$	$0.40, 0.77$	$0.40, 0.85$	$0.40, 0.88$	$0.40, 0.78$	$0.40, 0.51$

exceeded Rose's bid, but his shaded bid did not exceed Rose's bid, Colin bid 0.001 over Rose's bid. Finally, if Larry's valuation was above the current bid, Larry bid 0.001 over the current bid. Here, the payoffs to Rose and Colin are listed (in that order).

Of the strategies tested, Rose's strategy to shade by a factor of $2/3$ (shaded in gray) dominates all of her other strategies in the table, and Colin's strategy to shade by a factor of $1/2$ (also shaded in gray) dominates all of his other strategies in the table. This supports the earlier analysis. As a result, Rose and Colin will play toward the equilibrium with a payoff of 0.54 to Rose, 1.02 to Colin, and 2.59 to Larry (not shown in the table). From this, it is clear that the latter bidders in the auction have an advantage.

However, bids are restricted by the tiles available to each player. If players bid the largest sun tile available to them that is not above their shaded valuation, Rose has an average payoff of 1.23, Colin has an average payoff of 0.90, and Larry has an average payoff of 1.48, indicating the importance of the sun tiles.

7.3 ALL PAY AUCTIONS

In *Tammany Hall* (p.137), players take the role of individuals in Boss Tweed's Manhattan as they work to gain influence with different immigrant populations. Players bid this influence to win elections, and spent influence is lost whether the player wins or loses the election. Consider the five-player game of *Tammany Hall* from the start of the chapter. What is the appropriate bid for a bidder with a valuation of eight out of ten?

In an all-pay auction, players make bids, with the highest bid winning the auction and all players paying the value of their bid. As a result, bidding any positive amount could incur a loss without any benefit, which leads to a more conservative bidding strategy. Despite this difference, the same technique is used here as in the blind auction calculation. Like that situation, this finds an equilibrium, not a dominant strategy. The gain function is adjusted for a five-player game (so the probability factor is v^4). However, the critical change is that the gain function now has a negative term associated with losing the

auction,
$$g(v) = v^4(\theta - s(v)) + (1 - v^4)(-s(v)) = v^4\theta - s(v).$$

As before, differentiating g and setting $g'(\theta) = 0$ determines a differential equation for s, which is solved to obtain the solution.

$$0 = g'(\theta) = 4\theta^3\theta - s'(\theta)$$
$$s(\theta) = \frac{4}{5}\theta^5$$

So in the five-player game of *Tammany Hall*, with a uniform distribution between zero and ten, a player with a valuation of eight should bid $4/5\,(8/10)^5 \approx 0.26$ on the scale from zero to one, which corresponds to a bid of approximately 2.6 on the scale from zero to ten. The final result is the Optimal Bid in an All-Pay Auction.

Optimal Bid in an All-Pay Auction: In an n-bidder auction in which the valuations are independent and uniform on the interval $[0, 1]$, the Bayesian-Nash equilibrium strategy is to bid

$$s(\theta) = \frac{n - 1}{n}\theta^n$$

Since $0 \le \theta \le 1$, increasing the exponent of θ will reduce the bid, which matches the intuition that all-pay auction bids will be more conservative than blind auction bids.

All-pay auctions also appear in some combat mechanisms in tabletop games. In *Scythe* (p.123), players may exert power when entering combat. In these games, both sides lose all the power exerted, whether winning or losing the combat. In *Rising Sun* (p.125), players will bid to determine who takes which action, and all players will lose all of their bid coins. However, a twist in that game is that the winner distributes their bid among the losers. In *Nidavellir*, players recruit dwarves from a tavern and bid to determine who will select from the available dwarves first. All players will expend their money during the bid, but they will all recruit a dwarf. Higher bids result in recruiting better dwarves.

For Sale uses a similar mechanism. In this game, players take the role of a real estate broker. The game consists of two parts. A series of auctions will be held in the first part of the game. In an n-player game, n properties will be drawn and placed on the table. Players then bid to purchase these properties in a turn-based auction. However, when dropping out of the bidding, the player receives the remaining property with the lowest value and loses one-half of their most recent bid. In the second part, players bid to sell their properties (hopefully for a gain) in a blind auction. Because players in the first part of the game lose half of their bid, which is referred to as a partial-pay auction, the corresponding gain function (and the solution) are more interesting. Rather than solving this for the specific case of *For Sale*, a solution for the more general problem where losers pay a percentage, λ, of their bid will be developed. This solution provides a sliding auction method between the blind auction ($\lambda = 0$) and the all-pay auction ($\lambda = 1$). The auction in *For Sale* lands squarely in the middle ($\lambda = 1/2$).

The technique is the same as in previous work. The gain function, g, is

$$g(v) = v^{n-1}(\theta - s(v)) + (1 - v^{n-1})(-\lambda s(v))$$
$$= v^{n-1}\theta - (\lambda + (1 - \lambda)v^{n-1})s(v).$$

Setting $f(v) = (\lambda + (1 - \lambda)v^{n-1})s(v)$, this equation becomes $g(v) = v^{n-1}\theta - f(v)$. As before, differentiating and setting $g'(\theta) = 0$ will find an optimal solution for $s(\theta)$.

$$0 = g'(\theta) = (n - 1)\theta^{n-2}\theta - f'(\theta)$$
$$f'(\theta) = (n - 1)\theta^{n-1}$$
$$f(\theta) = \frac{n - 1}{n}\theta^n$$

Replacing $f(\theta)$ with its value and solving for $s(\theta)$ leads to the desired solution in the Optimal Bid in a Partial-Pay Auction.

Optimal Bid in a Partial-Pay Auction: In an n-bidder auction in which the valuations are independent and uniform on the interval $[0, 1]$, and where losers pay λ of their bid, the Bayesian-Nash equilibrium strategy is to bid

$$s(\theta) = \frac{n - 1}{n}\theta^n \cdot \frac{1}{\lambda + (1 - \lambda)\theta^{n-1}}$$

This result is consistent with both the Optimal Bid in a Blind Auction (when $\lambda = 0$) and the Optimal Bid in an All-Pay Auction (when $\lambda = 1$).

Figure 7.3 shows the optimal strategy functions in a four-player game with various values of λ. The extreme case where losers will pay twice their bid demonstrates that a positive bid is still warranted even in that extreme case. While losing the auction will result in a penalty, the long-term payoffs remain positive if one does not overbid, as a winning bid will bring a significant gain.

Before moving to the Revenue Equivalence Principle in the next section, it must be noted that all of the work in this chapter was predicated on independent private values. This is not the case in a situation where the item up for auction has the same value to all bidders. Such auctions are known as common value auctions.

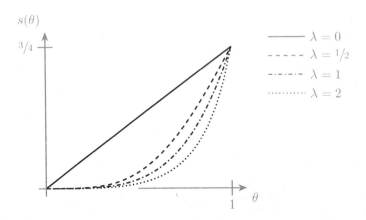

Figure 7.3: Bidding Strategies in Partial-Pay Auctions.

A common example of a common value auction is the dollar auction. This auction was designed by Martin Shubik[6] to demonstrate a paradox in rational bidding strategy. This is a two-bidder, ascending, all-pay auction for a one-dollar bill. In this case, once a bidder has made a bid and has been outbid, they are always incentivized to make a new bid. For example, if the first bidder has bid \$0.99, and the second bidder has bid \$1.00, the first bidder will compare not bidding again and losing \$0.99 to increasing the bid to \$1.01 and losing only \$0.01. Using the Optimal Bid in an All-Pay Auction, with a valuation of \$1, the first bidder should have bid only up to \$0.50. However, this assumes that both players have independent valuations of the dollar. More advanced work on auctions covers the analysis of common value auctions, see Appendix A.

7.4 REVENUE EQUIVALENCE PRINCIPLE

Given the variety of auctions available with different bidding strategies, is there an auction that is better for the seller?

The answer to this question is pursued assuming the players have independent, uniformly distributed private valuations. For n uniformly distributed values between 0 and 1, the kth-smallest value is expected to be $k/(n+1)$. These k values are known as order statistics and will be used without derivation here. So the lowest valuation is expected to be $\theta_1 = 1/(n+1)$ and the highest valuation is expected to be $\theta_n = n/(n+1)$. For the English auction (and its equivalent, the Vickrey auction), the amount paid will be the valuation of the second highest bidder. Under the assumptions about bidder valuations, the

---

[6]Martin Shubik was a twentieth-century American mathematician specializing in game theory and economics who developed the tabletop game *So Long Sucker* with Mel Hausner, John Nash, Lloyd Shapley.

expected revenue will be

$$E[\text{seller's revenue}] = \theta_{n-1} = \frac{n-1}{n+1}.$$

For the blind auction (and its equivalent, the Dutch auction), the amount paid will be a shaded bid from the highest bidder. The expected value of the highest bidder's valuation is $\theta_n = n/n + 1$, and the bid will be

$$E[\text{seller's revenue}] = \left(\frac{n-1}{n}\right)\theta_n = \left(\frac{n-1}{n}\right)\left(\frac{n}{n+1}\right) = \frac{n-1}{n+1}.$$

So, rational bidding in an English or blind auction will lead to the same revenue for the seller. The partial-pay auctions require more effort to analyze as they require that the probability distribution of the order statistics be known, something not covered in this book. However, in all of the partial-pay auctions, the expected outcome is again $(n-1)/(n+1)$. These results come from the Revenue Equivalence Principle, paraphrased here to avoid technical details.

Revenue Equivalence Principle: Under some reasonable conditions on bidder valuations and the form of the auction, any two auctions yield the same expected revenue to the seller.

CHAPTER 8

Logic

Figure 8.1: Components from *The Shipwreck Arcana*.[1]

The Shipwreck Arcana is a cooperative game. On a player's turn, they will hold two tiles numbered from one to seven in their hand, and have five cards with a placement restriction on the table. The Beast, shown in Figure 8.1, reads, "If one of your [tiles] is exactly 1 more or 1 less than the other, play one of them here." The player will place one of their tiles on one of the cards, and the other players will use the placement to attempt to determine the other tile.

Assume that Colin has tiles two and three. Which should be played in front of the Beast to give the other players the best chance of determining the other?

---

[1]The Shipwreck Arcana images and text were used with permission from Meromorph Games. They are copyrighted by Meromorph Games, and all rights are reserved worldwide.

 DOI: 10.1201/9781003383529-8

INTRODUCTION

Reasoning is an essential aspect of almost all tabletop games, as players must make inferences about the consequences of their actions. The entire book consists of examples of deduction, this chapter focuses on the use of deduction by the players to determine some hidden information.

In competitive deduction games, players work to determine the hidden information before any other player. In some games, like *Cryptid* (p.151), the other players will provide this information; in those situations, players will strive to give away as little information as possible. In other games, like *Turing Machine* (p.152) or *The Search for Planet X* (p.154), the game provides a component, an "oracle," which provides hints, rejects incorrect guesses, and confirms solutions.

In cooperative (and team) games, like *The Shipwreck Arcana*, players work together with all other players (or with their teammates in team games) to determine hidden information. The players giving clues, confirmations, and disconfirmations will strive to reveal as much information as possible, while the rules restrict the information that the players can convey.

8.1   COMPETITIVE DEDUCTIVE GAMES

When many people think of a deductive tabletop game, the first game that comes to mind may be *Clue*. In this game, players play characters trying to solve a murder mystery in a mansion. As part of this game, players must roll dice to move from room to room and may find themselves whisked to the other side of the manor against their will. This movement aspect of the game frustrates many players who are more interested in the deduction aspect of gameplay. Modern tabletop games provide the deductive mechanism while removing the roll-and-move aspect of the game *Clue*. One of the features of *Clue* is that it does not require an oracle to provide clues to the players. This is because other players will confirm or disconfirm conjectures using private information from the start of the game.

The game *Cryptid* has a similar deductive mechanism while removing the movement aspect of *Clue*. In *Cryptid*, players take on the roles of cryptozoologists searching for a cryptid in the American wilderness. The game board contains 108 spaces with various terrains (such as deserts, forests, mountains, or water) and some structures (such as abandoned shacks). At the start of the game, once the board is set up, each player receives a clue about the location of the cryptid. The types of clues available are known to all players. The clues have been carefully developed so that only one space on the board matches the clues of all players.

On Colin's turn, he may question Rose about the position of the cryptid by selecting a location and asking "Could the cryptid be here?" Rose will place one of her disks in that space if her rule allows the cryptid to be there, and she will place one of her cubes in that space if her rule does not allow the

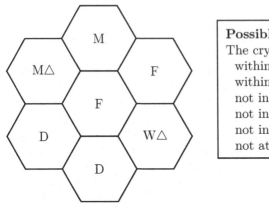

Possible clues:
The cryptid is ...
  within one space of a mountain.
  within one space of a desert.
  not in a mountain space.
  not in a desert space.
  not in a forest space.
  not at an abandoned shack.

Figure 8.2: Simple Terrain Layout from *Cryptid*.

cryptid to be there. An essential aspect of this game is that, on Colin's turn, if he forces Rose to place a cube on the board, Colin must also place one of his cubes in another space. Once a cube has been placed in a space, no other pieces may be placed there, and players can no longer make guesses about that location. This prevents players from placing cubes at a location already known to be impossible, so each cube placed provides new information. As the board fills up with cubes and disks, the cryptid's location gets increasingly limited until a player can determine its location (and win the game).

Figure 8.2 shows a seven-hex game version with terrains indicated by their first letter and abandoned shacks indicated by △. In a three-player game, each player is given a clue matching one of those shown at the right of the figure. Assume Colin has the clue "The cryptid is not in a desert space" and Rose has the clue: "The cryptid is not in a forest space." Colin has asked Rose if the cryptid is located in the northeastern forest tile. Rose must place a cube in the forest tile since it is a forest space. Colin learns that Rose's clue is either "within one space of a desert" or "not in a forest space." However, Colin must place a cube in one of the two desert spaces. If he places a cube in the southwestern desert, the only possible clue he could have would be "not in a desert space." However, if he places his cube in the southern desert space, he could have "within one space of a mountain" or "not in a desert space." It is, therefore, best for Colin to place a cube in the southern desert to keep the other players from determining his clue.

Using Venn Diagrams

In the game *Turing Machine*, players are attempting to deduce a three-digit code, where each digit is between one and five. The game provides from four to six verifiers to aid them in this task. Each verifier consists of a list of conditions, and the correct code passes precisely one of those conditions. For example, a straightforward verifier may have two conditions: "The first number is odd" and "The first number is even." At the start of the game, the player does not know which of these two conditions the verifier is using. However, the player may make proposals, and the verifier will inform the player whether the proposal passes the condition. So, suppose a player proposes one-two-three, and the verifier indicates that this proposal passes. In that case, the player learns that the verifier is testing the condition "The first number is odd." However, the player does *not* know that the first number is one because a three or a five also satisfies this condition.

Consider a reduced game of *Turing Machine* with only a two-digit code whose digits are between one and four (so 16 possible codes), together with the Three Verifiers in *Turing Machine*. On their turn, a player may verify their proposed code with two verifiers.

Three Verifiers in *Turing Machine*

Parity "Both numbers are even, or both are odd," or "One number is even, and the other is odd."

Threes "There are zero threes," "There is one three," or "There are two threes."

Smallest "The first number is the smallest," or "The second number is the smallest."

From these three verifiers, what can be inferred about the code? What codes should be proposed to determine the code?

Some inferences can be made even without proposing a code to be verified. The two digits in the solution cannot be the same because one of the two **Smallest** conditions must hold for the solution. This eliminates the four possible pairs and reduces the search to among 12 possible codes. This also eliminates the verifier "There are two threes" and leaves three pairs of possible verifiers, with one from each being the correct verifier. Here, a Venn diagram can be used to track which codes satisfy which properties, as shown in Figure 8.3.

This diagram places the possible codes in the locations based on their traits. This diagram shows two codes (32 and 43) that satisfy none of the listed properties and fall outside all three circles. Four of the regions have two possible codes. The game rules guarantee that the result of satisfying all three verifications will be unique. Codes in regions with more than one code cannot be the solution, as there would be no way to distinguish between them using a verifier. Using the terminology of Section 3.1, the two codes in those regions are equivalent. This reduces the possible codes to those in regions with a unique code.

Once the other codes have been removed from consideration, it is clear that using the **Parity** verifier does not provide any more information. Using the **Threes** verifier can distinguish between $\{13, 31\}$ and $\{24, 42\}$ and using the **Smaller** verifier can distinguish between $\{13, 24\}$ and $\{31, 42\}$. Using these

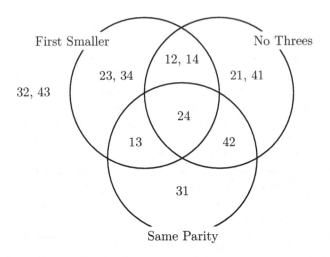

Figure 8.3: Venn Diagram of Possible Codes in *Turing Machine*.

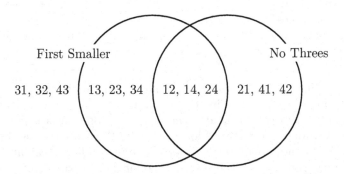

Figure 8.4: Venn Diagram of Possible Codes without the **Parity** Verifier in *Turing Machine*.

two verifiers will determine the code. This raises the question: is the **Parity** verifier relevant or redundant?

This can be answered by considering the Venn diagram without the **Parity** verifier. This Venn diagram is shown in Figure 8.4. Without the **Parity**, each code would be equivalent to two other codes, and no unique solution could be determined.

Using Trees

Another example of a competitive game that involves pure deduction is *The Search for Planet X*. In this game, players play astronomers searching for Planet X in one of twelve sky sectors, ten containing a single celestial object.

There are two comets, two gas clouds, one dwarf planet, four asteroids, and Planet X. Each object has specific placement requirements listed in Placement Restrictions for *The Search for Planet X*. The goal is to identify the sector containing Planet X and the objects in the adjacent sectors.

Placement Restrictions for *The Search for Planet X*
- Each Asteroid must be adjacent to at least one Asteroid.
- Each Comet must be in one of Sectors Two, Three, Five, Seven, or Eleven.
- The Dwarf planet and Planet X cannot be adjacent.
- Each Gas cloud must be adjacent to an Empty sector.

On a player's turn, they may take actions that provide them with clues as to the location of Planet X. For example, they can target a sector, and the oracle will tell the player what that sector contains with one exception: if the sector does contain Planet X, the player will be told that the sector appears empty. The player can survey for an object, allowing them to select a range of sectors and an object to be searched for. The player will learn how many objects of the type searched for are in the range of sectors. They can also research a topic that will provide them with an additional clue about the placement of the topic researched (for example, that at least one asteroid is adjacent to an empty sector). Some other game mechanics are included in the game that will hasten the end of the game (providing common knowledge to all players). To track this information, the game provides note sheets and suggestions on how to track the location of celestial objects. A sample note sheet is shown in Figure 8.5.

In this example, the player has determined some objects as shown in Figure 8.6 (where objects are abbreviated from their first letter). Sector Five cannot contain Planet X as Sector Four contains the dwarf planet, so Sector Five is truly Empty. How many more target actions must a player do to determine the location of Planet X and its neighbors?

A large component of these deduction games is efficiently tracking the consequences of hypotheses, and a rooted tree, as shown in Figure 8.7, can serve this purpose here. The vertices in the tree will be labeled with hypotheses about the location of the remaining objects using the notation "#X," where "#" is the sector and "X" is the first letter of the object. Here, the objects were considered in the following order:

1. the Empty sector was placed first (must be in 7 or 9),
2. the Gas cloud was placed next (must be next to Empty sector),
3. the Asteroids were placed next (must be together),
4. the Comet was placed third (must be in 2, 3, or 7),
5. Planet X was placed last.

Changing the order of the options will change the tree but not the outcomes. Vertices with a black interior represent branches that cannot be completed as the next object cannot be placed.

Figure 8.5: Components from *The Search for Planet X*.[2]

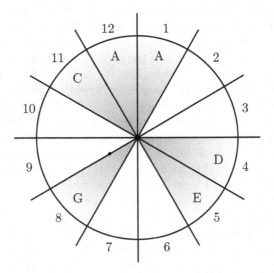

Figure 8.6: Partially Solved Puzzle in *The Search for Planet X*.

There are four possible placements; without any other information, it is reasonable to assume they are equally likely. Notice that at this point, when targeting a sector, it is possible to distinguish between an empty sector (which could only occur in Sectors Seven or Nine) and Planet X (which could only

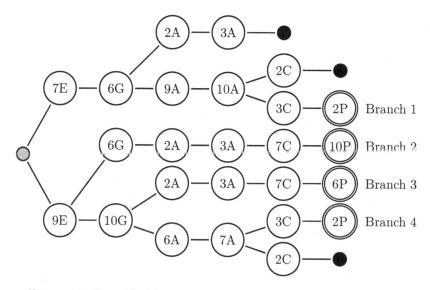

Figure 8.7: Possible Object Locations in *The Search for Planet X*.

occur in Sectors Two, Six, or Ten). Table 8.1 displays the possible branches based on information that can be obtained by targeting a sector. For example, the first row indicates that if an asteroid is found in Sector Two, then the result is either Branch Two or Three, and if Planet X is found in Sector Two, then the result is either Branch One or Four.

Notice that Planet X and its neighbors are the same in Branches 1 and 4, so these do not need to be distinguished to win the game. The shaded cells in the table indicate that a unique solution was discovered. The final column indicates the probability that targeting a sector will determine the location of Planet X and its neighbors. As seen in the table, all sectors other than Sector Nine have a 50% chance of determining the locations. Because so many branches occur with Sector Nine empty, targeting Sector Nine is less efficient.

Table 8.1: Possible Outcomes of Targeting a Sector in *The Search for Planet X*.

Target Sector	Object Discovered					Solution Percentage
	A	C	E	G	P	
2	2, 3				1, 4	50%
3	2, 3	1, 4				50%
6	4			1, 2	3	50%
7	4	2, 3	1			50%
9	1		2, 3, 4			25%
10	1			3, 4	2	50%

To answer the question, no single targeting can determine the location of all objects. However, one could determine the location of all objects by targeting two sectors (for example, Sectors Two and Ten).

8.2   COOPERATIVE DEDUCTION GAMES

Cooperative deduction games have all players working on the same side (or multiple players on the same team). In these games, the person providing the clue tries to provide clues that lead to correct guesses by their teammates. Examples of this type of game include *The Shipwreck Arcana* (p.150), *Hanabi* (p.161), *Codenames* (p.162), and *Letter Jam* (p.169). *The Shipwreck Arcana* will be discussed here, and the other three will be discussed in Chapter 9.

In *The Shipwreck Arcana*, the numbered tiles are referred to as fates, and the cards are called arcana cards. There are several arcana cards, each with text that restricts the fates' placement. Four cards are on the table each turn. Some of these arcana cards are shown in Four Arcana Cards from *The Shipwreck Arcana*, and these four cards will be assumed to be on the table. A fifth card is on the table, allowing the player to place a fate in front of it if they cannot meet other cards' placement requirements. All players will win if seven correct guesses are made before time runs out (time runs down as players choose not to guess or guess incorrectly). Colin and Rose are playing a game of *The Shipwreck Arcana*, with Colin holding Fate One and Fate Two. Which (if any) placements guarantee that Rose can correctly guess his fate?

Four Arcana Cards from *The Shipwreck Arcana*

Belltower ×3: If the sum of your fates is a multiple of 3, play one of them here.

Mirror Same: If both of your fates are the same, play one of them here.

Engine Factor: If one of your fates is double or triple the other, play one of them here.

Beast ±1: If one of your fates is exactly 1 more or 1 less than the other, play one of them here.

One way to analyze this is to consider the number of options that could result from each placement. Table 8.2 shows the available placement options for Colin.

Placing Fate One in front of the Beast would guarantee Rose a correct guess since placing Fate One in front of the Beast would require Colin to hold Fate Two. Now assume Colin holds Fates Three and Six. Which (if any) placements guarantee Rose a correct guess?

Again, consider the possible placements of Colin's tiles and the options afforded to Rose. Based on Table 8.3, there is no best option, as all placements lead to at least two possibilities for the fate in Colin's hand. However, the fact that the fate was *not* placed in front of the Mirror also provides information.

Table 8.2: Placement Options When Holding Fates One and Two in *The Ship-wreck Arcana*.

Card	Fate Placed	Options for Other Fate
Belltower	1	2, 5
Belltower	2	1, 4, 7
Mirror	—	—
Engine	1	2, 3
Engine	2	1, 4, 6
Beast	1	2
Beast	2	1, 3

Table 8.3: Placement Options When Holding Fates Three and Six in *The Shipwreck Arcana*.

Card	Fate Placed	Options for Other Fate
Belltower	3	3, 6
Belltower	6	3, 6
Mirror	—	—
Engine	3	1, 6
Engine	6	1, 2, 3
Beast	—	—

If Colin had a pair of matching fates, the Mirror would have allowed the other fate to be completely determined. So, by placing Fate Three in front of the Belltower, Colin indicates that he does *not* have a pair of matching fates, and the other fate must be Fate Six. Similarly, placing Fate Six in front of the Belltower leads to a single conclusion (namely that Colin is holding Fate Three).

Unlike the case of placing Fate One in front of the Beast, the conclusion derived from placing Fate Three in front of the Belltower is not forced by the rules of the game. It requires that Colin and Rose reason similarly about the game. To see that this can be problematic, consider the possible conclusion if Colin places Fate Two in front of the Beast. Rose knows that Colin holds either Fate One or Fate Three. Based on the reasoning associated with Table 8.2, if Colin held Fates One and Two, he could have guaranteed a correct guess by placing Fate One in front of the Beast instead of placing Fate Two in front of the Beast. Therefore, Rose may reasonably assume that Colin does not hold Fate One. Hence, placing Fate Two in front of the Beast signals that Colin holds Fate Three. Similarly, placing Fate Three in front of the Beast would indicate that Colin has Fate Four. Surprisingly, when Colin places Fate Three in front of the Beast, Rose guesses that Colin holds Fate Four, which is revealed to be incorrect. Did Colin misplay his fates?

Not necessarily. Colin could have reasoned: "If a player places Fate Seven in front of the Beast, they must have a Fate Six. Therefore, if a player has Fates Six and Seven, they will place Fate Seven in front of the Beast. So, if a player places Fate Six in front of the Beast, it must signal that the player has Fate Five. Placing Fate Five would signal that the player has Fate Four, placing Fate Four would signal that the player has Fate Three, and placing Fate Three would signal that the player has Fate Two. Therefore, if I place Fate Three in front of the Beast, Rose will know I hold Fate Two."

The situation in *The Shipwreck Arcana* is similar to the difference between a dominant strategy and a Nash Equilibrium. A dominant strategy is a strategy that is optimal, regardless of the opponent's reasoning. For example, placing Fate One in front of the Beast only leaves one option for Rose by the game's rules. On the other hand, playing toward a Nash Equilibrium is only appropriate if the other player is also reasoning the same way. In this case, placing Fate Three in front of the Beast only leaves one option *if* Rose reasons in the same way Colin has reasoned.

CHAPTER 9

Number Theory

Figure 9.1: Components from *Hanabi*.[1]

Hanabi is a cooperative card game, where players attempt to create a spectacular fireworks show. The cards each have one of five colors and one of five ranks. The goal is to place the cards in ascending stacks, one for each of the five colors, as shown in Figure 9.1. This task is challenging as players cannot see their cards but can see the other players' cards. On their turn, a player may either play a card from their hand into a stack or provide information to another player. When providing information, a player can either state a color and point at all cards of that color in one other player's hand *or* state a number and point at all cards of that rank in one other player's hand.

In *Hanabi*, how much information can Colin give with a single clue on his turn?

---

[1]©2023 R&R Games, Inc.

DOI: 10.1201/9781003383529-9

INTRODUCTION

This section continues the theme of cooperative deduction games but is more focused on circumventing the restrictions of the game rather than abiding by them. It should be noted that this whole chapter is about bending (or breaking) the rules in a cooperative game to increase the chance of winning. The result of employing any of these strategies removes much of the fun from the game and is not intended for use in serious play, where the goal is to have fun with friends.

However, when presented with rules restricting options, a natural question is how far they can be bent using mathematical techniques to compress the information to be conveyed. In this chapter, information will be encoded using prime factorizations, binary representations, and modular arithmetic.

9.1 THE FUNDAMENTAL THEOREM OF ARITHMETIC

In *Codenames*, the players are divided into two teams, and one player on each team is selected as the spymaster. Then, a five-by-five rectangular grid of cards is laid out, each card displaying one word. The spymaster is given a key that identifies eight or nine grid locations (depending on whether their team is playing first). Assume that the spymaster is on the team that plays first and must identify nine locations out of the 25 locations in the grid. The goal of the spymaster is to provide clues to their teammates that will allow them to select the locations identified on the key. Each clue consists of a single word and an integer. The word should relate to words in identified positions, and the integer should indicate the number of positions the word identifies. So, for example, if two locations marked on the key contain the words "Bat" and "Bird," then the spymaster might choose the clue of "Winged: Two" to indicate that there are two words on the table related to being winged that mark locations identified on the key. Two rules about clue-giving are that the word must be related to the words in the grid and that the integer portion of the clue cannot be used to identify words. The example in the rule book states that the number "eight" cannot be used as the integer part to hint that "octopus" is one of the identified words. How much extra information can be provided in a single clue if the word and number are allowed to be anything?

At first, it may appear that, at best, the integer part of the clue might give one extra location, maybe two if the spymaster were lucky, or none if the spymaster were unlucky. However, infinite information can now be transmitted with the clue, which is more than enough to allow all identified cards to be selected with a single clue. This is because the Fundamental Theorem of Arithmetic can encode an infinite amount of information into a single integer.

Fundamental Theorem of Arithmetic: Any integer n with $n > 1$ can be written as the product of powers of ascending primes. Furthermore, the factors in this product are unique.

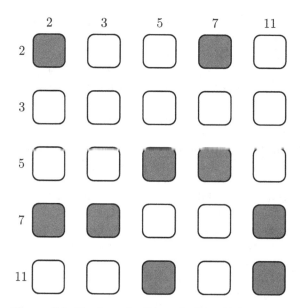

Figure 9.2: Cards to be Identified in *Codenames*.

To encode the location of every card using the Fundamental Theorem of Arithmetic, the first step is to establish a convention that encodes each location into an integer such that the product of these integers preserves all of the original information. One way to do this is to assign the rows and columns to prime numbers. The first grid row is assigned the value two, the second grid row is assigned the value three, the third is assigned the value five, the fourth is assigned the value seven, and the fifth is assigned the value eleven. The essential aspect of this is that these numbers are all prime. Make the analogous assignments for the columns. Now, on each row, multiply all of the column values associated with the cards to be identified and place this number into the exponent of the row's value. Then multiply the row products to arrive at the integer to give in the clue.

To make the process a little less confusing, consider the situation in Figure 9.2, where the cards to be identified are shaded.

The first row consists of cards in the first and fourth columns and therefore is encoded as $2^{2\times7} = 2^{14}$. The second row does not contain any cards and is encoded as $3^1 = 3$ (remember from Section 1.2, that empty products evaluate to one). Continuing, the third row results in 5^{35}, the fourth row results in 7^{66} and the fifth row is 11^{55}. Multiplying all of these together produces a number roughly equal to 1.6×10^{142}. However, there is no stipulation as to how the number needs to be expressed, so this number would likely be presented in its factored form, giving the clue

$$\text{kumquat: } 2^{14} \cdot 3^1 \cdot 5^{35} \cdot 7^{66} \cdot 11^{55}.$$

Players are free to choose any word they wish. The word kumquat was chosen because it sounds funny. As noted earlier, this is an absurd way to play the game. However, the idea of encoding information in products of prime powers was used to prove Gödel's Incompleteness Theorem.[2] The proof assigned symbols used in mathematical proofs to prime numbers and sequences of symbols to products of sequential primes. For example, Gödel encoded the symbol "0" as the integer one, the symbol "=" as the integer five and the equation "0 = 0" as $2^1 \times 3^5 \times 5^1 = 2430$.

While the method is undoubtedly clever, it requires the use of numbers up to $11^{2 \cdot 3 \cdot 5 \cdot 7 \cdot 11} \times 7^{3 \cdot 5 \cdot 7 \cdot 11} \approx 5 \times 10^{3381}$ while skipping many numbers which could never be used to identify nine cards. Combining this with the difficulty of factoring an integer, a difficulty that grows exponentially relative to the integer's size, this method is not very attractive.

Once the idea of encoding locations into integers is discovered, more efficient methods can be explored. A slightly more compact method encodes the information in a sequence of binary digits, where the grid is traversed from left to right and then down. A one is used for an identified location, and a zero is used for an unidentified location. The example in Figure 9.2 would be encoded as 100100000000101100100101. In this case, the largest required number would be 111111110000000000000000. Interpreting these as binary numbers, the largest integer required would be

$$\sum_{k=16}^{24} 2^k = 33\,488\,896.$$

This is significantly smaller than the value required by the previous method. This raises the final question for this section: what is the smallest number of integers required to encode this information? This would be equal to the number of ways to select nine objects from a collection of 25 objects, which is $C(25, 9) = 2\,042\,975$.

9.2 MODULAR ARITHMETIC

The strategy for giving information in *Hanabi* (p.161) is based on a different principle and will be introduced with a collection of games collectively known as "hat guessing games" in the mathematics community. It should be noted that mathematicians, not hobby gamers, play these games, but they introduce the basic ideas that will be applied in more complicated situations.

A surprising amount of mathematical literature focuses on gnomes guessing the colors of their hats (to be honest, even one article would be surprising). These hat guessing games involve a group of gnomes who are wearing colored

---

[2]Kurt Gödel was a twentieth-century Austrian-American mathematician whose Incompleteness Theorem proved that any axiomatic system that can describe the integers must either be inconsistent (meaning it can prove some false statements) or incomplete (meaning it cannot prove some true statements).

hats. These gnomes cannot see their hat but can see the other gnomes' hats. Each gnome will be required to guess the color of their hat. The goal is to find a strategy that either maximizes the expected number of correct guesses or minimizes the number of incorrect guesses.

Two gnomes (say Anders and Birgitta) are in a room together, each with either a red hat or a blue hat. The goal is for at least one of the two gnomes to guess the color of their hat (or, equivalently, for one of the gnomes to be able to determine the hat color of every gnome, including themselves). From Chapter 1, it can be determined that there are $2^2 - 4$ possible hat distributions, but the gnome's guesses will only cover two of these combinations. So random guessing will not always succeed, and using the techniques from Section 5.1, it can be determined that random guessing will result in success only 75% of the time. However, if Anders and Birgitta can develop a strategy before being required to guess, they can do much better by changing their perspective on the problem. Instead of trying to determine the color of every hat (four possible configurations), they can try to determine if their hats are the same color or different colors (two possible configurations). Under this strategy, one gnome (say Anders) will be selected to make a guess consistent with the hats being the same color, and the other gnome will make a guess consistent with the hats being different colors. When Anders makes his guess, he will look at Birgitta's hat and then guess that color. When Birgitta makes her guess, she will look at Anders' hat and then guess the other color. This will guarantee that one gnome is correct (but it also guarantees that one gnome is incorrect). Is there any way to implement a similar strategy for six gnomes and five hat colors?

The short answer is yes. However, understanding the strategy requires understanding modular arithmetic. Modular arithmetic is not a single system of arithmetic. Instead, it refers to several related systems, one for each positive integer. The positive integer on which the system is based is called the modulus of the system. The phrase "modulo n" indicates that the calculation uses a modulus of n.

Modular arithmetic is a finite version of ordinary integer arithmetic. It is used in everyday life: when recording times, Americans use a mod twelve system for hours. It is also used in modern computer science: modular arithmetic is essential to secure internet connections. This section focuses on modular addition, subtraction, and multiplication, as they are all very similar and will be sufficient for the purpose here. Some modulus values also support division, which is more complicated and unnecessary for the applications here.

Two integers, p and q, are equivalent modulo n, written $p = q \pmod{n}$, if their difference is an integer multiple of the modulus, n. So, working modulo five,

$$\ldots \equiv -9 \equiv -4 \equiv 1 \equiv 6 \equiv 11 \equiv \ldots \pmod{5}.$$

The terminology and notation indicate that while -9, -4, 1, 6, and 11 are not identical integers, they will have all act the same when operated on modulo five. This notion of equivalence satisfies the properties of an equivalence relation (see Section 3.1). It is analogous to the situation in that section, where two different tiles were equivalent if one could be rotated or reflected to match the other.

For each modulus, there are corresponding operations of addition, subtraction, and multiplication (using the notation $\oplus_n$, $\ominus_n$, and $\otimes_n$ for the operations with modulus n). Performing the modular operation begins as normal: add, subtract, or multiply the two numbers using the standard rules for integer arithmetic. However, modular arithmetic adds one more step: translating this value into the range between 0 and $n-1$ (inclusive) by adding or subtracting multiples of n. Continuing with the modulo five example,

$$3 + 4 = 7 \equiv 2 \pmod 5 \quad \text{so,} \quad 3 \oplus_5 4 = 2$$
$$8 + 7 = 15 \equiv 0 \pmod 5 \quad \text{so,} \quad 8 \oplus_5 7 = 0$$
$$3 - 4 = -1 \equiv 4 \pmod 5 \quad \text{so,} \quad 3 \ominus_5 4 = 4$$
$$3 \times 4 = 12 \equiv 2 \pmod 5 \quad \text{so,} \quad 3 \otimes_5 4 = 2$$

Back to the Gnomes

Modular arithmetic can be used to generalize the case of Anders and Birgitta to an arbitrary number of gnomes and hat colors. Instead of two gnomes, consider a room with six gnomes; each is wearing a hat with one of five possible colors. With $5^6 = 15\,625$ hat configurations possible, random guessing leads to a probability of success of $6/15625$ or roughly 384 successes for every million attempts. However, there is a strategy that the gnomes can use to guarantee at least one correct guess. Since the goal is to determine the hat color and there are five colors, each color will be assigned a number between zero and four, and the work will be done modulo five. Assign each gnome a number. For example, the gnomes could be numbered sequentially, starting at one. Each gnome will compute the sum of the numbers associated with the hats they see. They will then subtract this value from their assigned number (all operations computed with modulus five). For example, Table 9.1 shows this for one distribution of hat colors.

Table 9.1: Hat Color Distribution Example.

Gnome	Hat Color	Sum	Guess
1	1	$3 \oplus_5 1 \oplus_5 2 \oplus_5 0 \oplus_5 1 = 2$	$1 \ominus_5 2 = 4$
2	3	$1 \oplus_5 1 \oplus_5 2 \oplus_5 0 \oplus_5 1 = 0$	$2 \ominus_5 0 = 2$
3	1	$1 \oplus_5 3 \oplus_5 2 \oplus_5 0 \oplus_5 1 = 2$	$3 \ominus_5 2 = 1$
4	2	$1 \oplus_5 3 \oplus_5 1 \oplus_5 0 \oplus_5 1 = 1$	$4 \ominus_5 1 = 3$
5	0	$1 \oplus_5 3 \oplus_5 1 \oplus_5 2 \oplus_5 1 = 3$	$5 \ominus_5 3 = 2$
6	1	$1 \oplus_5 3 \oplus_5 1 \oplus_5 2 \oplus_5 0 = 2$	$6 \ominus_5 2 = 4$

From this, the gnome assigned the number three got a correct guess, but were they just lucky? When presented as an algorithm with no explanation, it is unclear whether this technique will always work. One of the goals of mathematics is to verify that techniques always succeed (or to be able to list every case where the technique may fail). The above strategy will always work because instead of requiring the gnomes to determine all of the hat colors (of which there are 15 625 configurations), it only attempts to determine the modular sum of the hat colors (and there are only five such values). Each gnome is essentially guessing that the modular sum equals their assigned number. So, a gnome will guess correctly if the sum of hat colors in the room matches their assigned number. Because there are at least as many gnomes as colors, every possible modular sum can be assigned to at least one gnome, and that gnome will make a correct guess. If there are more gnomes than colors, there is even the possibility of multiple gnomes with a correct guess. If there are fewer gnomes than colors, one can still use this strategy to maximize the probability that at least one of the guesses is correct. This strategy is a generalization of the Anders and Birgitta strategy earlier. In the case of two hat colors, the modulus would be two, and if the hats were the same color, they would sum to zero (and Anders would be correct), and if they were different colors, they would sum to one (and Birgitta would be correct).

Using a Signal

If the gnomes make guesses one at a time and each gnome knows the previous guesses, then they can do much better. Returning to the two-gnome case where the gnomes decide that Anders will guess first. In this case, he and Birgitta can agree that he will guess the color he sees on Birgitta's head (sending a signal to Birgitta about the color of her hat), then Birgitta should guess the same color that Anders guessed. In the original strategy, one guess would be correct, and one guess would be incorrect. In this strategy, one guess is correct (Birgitta will always guess correctly), but there is also a 50% chance that Anders is also correct. Notice that this also allows for more than two colors: no matter how many colors the hats could be, Anders can guess Birgitta's hat color, and Birgitta will guess the same color and be correct.

This technique extends to any number of gnomes and colors. One of the gnomes will be elected to be the signaler, and they will make the first guess (the signal) by guessing that their hat has the color associated with the *sum of the hats of the other gnomes*. By broadcasting this signal, the signaler will allow all of the other gnomes to guess the color of their hat correctly. Each gnome does this by computing the sum of all the hats they and the signaler can see. They can subtract this number from the signal to determine their hat's color. The value obtained this way is that gnome's message. Notice that while every gnome receives the same signal, different gnomes can receive different messages. The process of sending multiple messages using a single signal is referred to as multiplexing (or muxing for short), and the reverse of

Table 9.2: Sending a Signal: Gnome 1 Sends the Signal $3 \oplus_5 1 \oplus_5 2 \oplus_5 0 \oplus_5 1 = 2$.

Gnome	Hat Color	Sum	Message
2	3	$1 \oplus_5 2 \oplus_5 0 \oplus_5 1 = 4$	$2 \ominus_5 4 = 3$
3	1	$3 \oplus_5 2 \oplus_5 0 \oplus_5 1 = 1$	$2 \ominus_5 1 = 1$
4	2	$3 \oplus_5 1 \oplus_5 0 \oplus_5 1 = 0$	$2 \ominus_5 0 = 2$
5	0	$3 \oplus_5 1 \oplus_5 2 \oplus_5 1 = 2$	$2 \ominus_5 2 = 0$
6	1	$3 \oplus_5 1 \oplus_5 2 \oplus_5 0 = 1$	$2 \ominus_5 1 = 1$

this process is referred to as demuxing. The case above, with the first gnome selected as the signaler, is presented in Table 9.2, where all gnomes except the signaler make correct guesses.

This process works because modular arithmetic shares many properties of ordinary integer arithmetic. The property that is being used here is that for every a and b between 0 and $n - 1$, there exists a unique value of x between 0 and $n - 1$ which solves the equation $a \oplus_n x = b$.

The Group $\mathbf{Z_n}$

The remainder of this section is a little more technical and ties modular arithmetic to groups (from Chapter 3.1). Recall that a group was defined as a collection of actions that satisfy the Group Properties on page 49. An operation is associative if $a(bc) = (ab)c$ for all values of a, b, and c. Generally, a group is any collection of objects with an associative operation that satisfies the Group Properties.

In Chapter 3, the collection consisted of symmetries of a shape, while in this chapter, the set consists of numbers. In Chapter 3, the operation was the chaining of actions, while in this chapter, the operation is modular addition. For each positive integer, n, the group $\mathbf{Z_n}$ is defined to have the elements $\{0, 1, 2, \ldots, n-1\}$ and to have the operation $\oplus_n$. The element 0 is the identity element, and the inverse of x is the number $0 \ominus_n x$. The choice of the notation $\mathbf{Z_n}$ comes from the German word "Zählen" which refers to counting. The integers are denoted by $\mathbf{Z}$ and the subscript, n, in $\mathbf{Z_n}$ indicates arithmetic modulo n.

A group that has the same structure as $\mathbf{Z_n}$ is Rot_n, the set of rotations of the regular n-gon. Each element of $\mathbf{Z_n}$ has a corresponding element of Rot_n, with the number k corresponding to k clockwise rotations through an angle of $\alpha = 360°/n$. Furthermore, this correspondence "preserves the group structure." This means that doing k rotations of α followed by doing ℓ rotations of α (the operation in Rot_n) corresponds to doing $k + \ell$ rotations of α (where the operation was done in $\mathbf{Z_n}$). Recognizing that two structures are equivalent is one of the goals of mathematics. In the context of groups, $\mathbf{Z_n}$ and Rot_n are equivalent.

Just being equivalent groups does not mean that they are identical. For example, $\mathbf{Z_n}$ has more structure than Rot_n because there is an additional operation, multiplication, in $\mathbf{Z_n}$ that does not have a counterpart in Rot_n. For example, $3 \times 4 = 2$ has meaning in $\mathbf{Z_5}$. However, 3, 4 and 2 in $\mathbf{Z_5}$ corresponds to R_{216}, R_{288}, and R_{144} in Rot_5 and it is unclear whether there is any natural operation that would give meaning to $R_{216} \times R_{288} = R_{144}$.

9.3 MUXING IN GAMES

Leaving the gnomes behind, this section presents a situation where muxing can allow near-optimal gameplay in a tabletop game. *Letter Jam* is a cooperative word game for two to six players. The standard difficulty game with six players will be analyzed here. The primary game component is a deck of 64 cards, each card has one of 21 letters on its face (JQVXZ are not used). At the start of the game, each player receives a sequence of five facedown cards (this analysis will assume these are random). Players are not allowed to look at the faces of their cards. Instead, each player sets their leftmost card on a stand with the card back toward them so that the card's face is visible to everyone else. In casual play, the game's goal is to allow every player to guess all their letters in a fixed number of rounds (for six players, there are eleven rounds). There is also a scoring system available, and the strategy described here can also be used to maximize scores.

A round of the game has one player giving a clue by spelling a word using the letters that they can see and (possibly) a wildcard letter, denoted by *. At the start of each round, the players determine who will give a clue that round by consensus, often following a discussion where players make statements about the length of their proposed clue and how many player letters will be used. For example, suppose Colin can see the five letters: B, L, C, N, and O in clockwise order starting with the character to their left. Colin may consider using the word JOB and say he can create a three-letter word using letters from two different players and using the wildcard.

The clues are given by placing numbered tokens in front of other players' letters to spell a word (a card with * is placed in the center of the table for the wildcard). Any letter can be used more than once with the condition that the wildcard letter (if used) must always be used for the same letter throughout the round. The expected strategy is that each player whose letter was used in the clue would try to use their vocabulary knowledge to deduce the letter in front of them based on the other letters in the word spelled by the clue-giver.

For example, if Colin is selected to give the clue, he would put token one in front of *, token two in front of O, and token three in front of B. Each player can then write out the letters of the clue, using a ? to indicate their letter (which they are trying to guess). For example, the player with the O knows the word has the pattern *?B. If their vocabulary is sufficient, they may recognize that only 57 possible three-letter words end in B. Looking at the second letter of each word, the ? must be one of only eight characters. Similarly, the player with the B has 239 possibilities for *o? and ? must be one of 24 letters, ruling out only J and Q. So, no new information has been provided to the player with the B, as J and Q are not included in the game. This clue also provides no information to the players with the L, C, or N.

If Colin had used the word BALLOON (with the wildcard for the A), more information would have been provided to the other players. The player with the B would have two options (B and G), the player with the L would have three options (L, S, F), and the players with the O and the N would each have only one option.

If a player believes they know their revealed letter, they turn their revealed card face down and place the next letter on their stand. Because the player giving the clue cannot be expected to gain any information about their letter, the shortest a six-player game can last is six rounds (each player completing one round as a clue-giver and one round receiving a clue for each of their five cards).

Only the person giving a clue needs to know the word used as a clue, but in traditional play, if no one else knows the word, the clue would be wasted. The joy in *Letter Jam*, as in *Codenames* (p.162), arises in the clever vocabulary choices that provide as much information as possible to as many players as possible. Words must be general enough that other players will recognize them even with some missing characters and must be unique enough that it limits the number of possible missing letters. Some groups focus on finding long words that involve as many players as possible (for example, choosing BALLOON over JOB).

Muxing Letter Jam

A round of *Letter Jam* is precisely the type of problem that the methods of the previous section can solve. Each player can see the other players' letters but cannot see their own. To use the gnome strategy, the players must agree on a convention to assign the numbers from 0 to 20 to the 21 possible letters (the matching of A to zero, B to one, etc. is the one used here). The choice of messages is clear: each player should receive the number associated with their letter. However, the game constrains what signals are allowed, so using the strategy effectively requires a way to send the 21 signals.

The signal will consist of the placement of the numbered tokens. For now, assume that the clue-giver has sufficient vocabulary to place any token before any player. Placing a single token allows for five signals in a six-player game (clue-givers are not allowed to place a token in front of themselves). This is insufficient to send 21 different signals. However, using two tokens provides 25 different signals, which is sufficient. For each round, assign numbers to the players clockwise, starting with zero for the player to the left of the clue-giver. This means that the player to the right of the clue-giver is numbered four.

The first two tokens placed in front of players will construct the signal. The first token will indicate the number of fives in the signal, and the second will indicate the number of ones in the signal. For example, the clue-giver would indicate the number $14 = 2 \times 5 + 4 \times 1$ by placing the first token in front of Player Two and the next in front of Player Four. Tokens placed in front of the wildcard character are ignored.

Continuing the example, Colin can see B, L, C, N, O and must send the signal of 20:

$$\underbrace{1}_{B} \oplus_{21} \underbrace{11}_{L} \oplus_{21} \underbrace{2}_{C} \oplus_{21} \underbrace{13}_{N} \oplus_{21} \underbrace{14}_{O} = 20.$$

The number 20 has four full 5s with no left-over 1s, so the first token must go to Player Four (the player with the O), and the second token must go to Player Zero (the player with the B). The word JOB will work (as tokens in front of the wildcard are ignored).

Just like the situation with the gnomes, each player can now determine their letter. For example, the player with the C computes the sum of the letters both they and the clue-giver can see: $1 \oplus_{21} 11 \oplus_{21} 13 \oplus_{21} 14 = 18$. Knowing that the sum that the clue-giver can see is a 20, they compute $20 \ominus_{21} 18 = 2$ and now know that they have a C.

Using this strategy, if every player can always give a clue, the players can determine five of the six letters each round and complete the game in six rounds (with everyone giving exactly one clue).

The English Language

However, it violates the rules if the clue-giver does not know a word that could be made with the letters used in the clue. Even with the wildcard, are any combinations impossible to use as a signal?

A dictionary search indicates that every two-letter pair allows for constructing at least one word except for the following five pairs: FC, FK, GC, NC, and UW. This does not mean that there are no words with these pairs. For example, FACE could be used for FC if the wildcard were used for A and another player had E. However, the goal is to have a method that works in all cases. One solution might be to avoid the problematic letters. Earlier, it was assumed that the five letters in front of each player were random, but this is not how the game starts. Each player uses a collection of cards to create a five letter word, which they pass to the other player on their right to form that player's unrevealed letters. Players can agree not to select a W, C, or K in these initial words, and this will guarantee that every required signal can be sent (as well as reduce the number of required signals from 21 to 18). If players cannot select the letters used, another solution would be to skip a player's turn as a clue-giver if they cannot give a clue using this convention.

9.4   MUXING WITH SEVERELY RESTRICTED SIGNALS

As in *Letter Jam* (p.169), *Hanabi* (p.161) allows each player to know the contents of every other player's hand, but not the contents of their own hand. It might be hoped that muxing can be used to attempt to communicate here as well. To implement a muxing strategy, one needs to determine two things: the messages to send and how to send signals using the game's mechanisms. In the case of *Letter Jam* (p.169), the messages were clear, the unrevealed letter, and one was able to use the placement of tokens to generate the needed signals. The communication restrictions in *Hanabi* are more severe, so the first step is to determine how many signals are available.

In *Hanabi*, the clue-giver must select one player and give that player one clue. The clue can either identify all cards of a particular suit in that player's hand or all the cards of a particular rank. It is explicitly stated in the rules that a player cannot identify a suit or rank that the player does not have in their hand. This means that the only information that can be used for the signal is which player the clue is given to and which type of clue (suit or rank) is given. Answering the opening question, in a five-player game of *Hanabi*, only $4 \times 2 = 8$ signals may be sent. If one were allowed to provide clues that a particular color or rank was not in a player's hand, then a signaler could give any of five rank clues and five color clues for a total of $5 + 5 = 10$ clues to any of the other four players for a total of $10 \times 4 = 40$ clues, a five-fold increase in the amount of information.

In *Letter Jam*, there were enough signals that a message that identified all of the hidden information for each player could be sent. In *Hanabi*, there are 50 different cards (with some duplicates), and each player has a hand of four cards. Careful counting using generating functions indicates there are 18 480 possible hands. With only eight signals, it is unclear what signals will be the most effective. Two strategies were presented in "How to Make the Perfect Fireworks Display: Two Strategies for Hanabi" [21] and compared to the **Cheating Strategy**, where each player knows the contents of their hand.

The **Recommend Strategy** focuses on messaging whether one recommends that a particular card be played or discarded. The message to a player would indicate a card to play (four options) or a card to discard (four options) for eight possible messages.

The **Information Strategy** focuses on messaging information about the contents of each player's hand by targeting the most critical card in each player's hand and trying to allow the player to identify the rank and suit of that card. The technique is relatively involved, and interested readers should consult the original article.

A perfect score in *Hanabi* is 25 points (for creating a stack of cards numbered one to five for each of the five colors). In simulations, the **Recommend Strategy** resulted in perfect scores 16% of the time with a median score of 24, the **Information Strategy** resulted in perfect scores 76% of the time, and the **Cheating Strategy** resulted in perfect scores 91% of the time.

Resources

Having made it to the Appendix, the next natural question is: "Where to go now?"

The background resources used when writing this book are collected here. The first section will focus on general books covering many topics, followed by sections related to the topics in this book. Readers should also check the website at http://www.monsterworks.com/ludi/ for additional material developed with this book, including additional content and computer code in *Mathematica* and Python for many examples. Feel free to contact me at magister_ludi@monsterworks.com with suggestions for other resources or topics to cover (or to ask to join me for a game).

General Resources

The mathematics textbooks referred to when covering these topics are listed in the table below. If you need help figuring out where to start, consider these. The table below indicates which books cover which topics. In addition to these books, *Teaching Mathematics Through Games* [16] includes many topics from this book but is focused on using games in the classroom (and so is probably of most interest to teachers).

Resource	This Book's Chapters								
	1	2	3	4	5	6	7	8	9
Groups and Symmetry [5]			•						
Introductory Combinatorics [12]	•		•	•					•
Networks, Crowds, and Markets [24]					•		•	•	
Combinatorics and Graph Theory [32]	•		•	•					
Mathematics for Computer Science [39]	•			•	•			•	•
Game Theory: An Introduction [47]							•	•	
Games, Gambling, and Probability [49]	•				•				

DOI: 10.1201/9781003383529-A

Several books discuss the games that mathematicians play. These books present classical and common combinatorial games, like *Hex* (p.118). Examples of these texts are *The Mathematics of Games* [8], *Luck, Logic, and White Lies* [10], *Math Games with Bad Drawings* [42], and *Mathematicians Playing Games* [37]. The second is notable for its extensive bibliography, and the third for its copious stick-figure illustrations.

An excellent place to start on the gaming side is *Gametek: The Math and Science of Gaming* [25], which includes chapters on game theory, math, and other exciting topics outside the scope of this book, such as psychology, game mechanics, and history. *Achievement Relocked: Loss Aversion and Game Design* [26] focuses more on psychology. However, it is worth noting that tabletop play is as much (if not more) a result of psychology as mathematics. Sam Mcdonald, the designer of *Architects of the West Kingdom*, expressed this sentiment in a recent Board Game BBQ podcast episode.

> "The other thing that I've learned is that a game can be mathematically balanced, like really mathematically balanced, but if it doesn't feel balanced, people will say it is not balanced.... If someone says that something is not mathematically balanced and I show them a spreadsheet, I'm not sure that's going to convince them, I think the thing that is going to convince them is the experience."
>
> —Sam Mcdonald [40]

As the preface mentions, *Eurogames: The Design, Culture and Play of Modern Board Games* [56] is an excellent introduction to the tabletop game hobby. You can learn more about how a tabletop game is played (including all the rules omitted in this book) through online sites. Many publishers have posted these games' rulebooks online (allowing you to read through the rules), or, perhaps of more use if you are not planning on playing the game, you can find "How to Play" videos and play-through videos on sites such as YouTube which demonstrate how the games are played. In addition to videos, one can find a wide variety of podcasts that discuss tabletop games. Some links to videos and podcasts are provided at the monsterworks site referenced above.

A large community centered on tabletop gaming can be found at BoardGameGeek, https://www.boardgamegeek.com, often referred to as "The Geek." This website can be overwhelming to new users but does contain a wealth of information about tabletop games once one learns to navigate it. You can search for a particular game or browse all games. Of particular note are its crowd-sourced ranking (listed as the game's "Average Rating") and the site's Bayesian average of these rankings (listed as the game's "Geek Rating") that identifies games with many high ratings. The other crowd-sourced ranking on the site is the game's "Weight," which refers to the game's complexity. While there is frequent debate about the accuracy of this value, it can provide some level of guidance. If you have only played games with a weight of two or lower and are now considering a game with a weight of three or higher, you should expect it to take longer to learn than your previous fare. To give a rough idea of the range, *Candy Land* has a weight of 1.10; *Ticket To Ride*, *Carcassonne*, and *Catan* all have weights near 2.00; while the games with the 50+ page rulebooks mentioned in the Preface have weights above 4.25. User accounts are free; users can post on a game's forum requesting help with rules. In some places, the game's designers may answer, but more often, answers will come from other enthusiasts.

If you are interested in playing tabletop games but are concerned about the cost (in terms of either money or storage space), consider online gaming websites. Several services provide access to some of the games referenced in this book (membership is usually free, with some games requiring a paid subscription). Links to some of these sites are posted on the `monsterworks` website.

Sources by Topic, in order of appearance

Polygonal Tilings: The paper "Tilings by Regular Polygons" [31] considers tilings by regular polygons but does not restrict the tiling to only a single polygon type.

Regular Polygons: The dearth of triangular tilings was noted by *Triangular Grids - there are almost no games that use it. Only hex and square grids. Have you heard of any games using triangles, do you have experience of playing/developing these?* [52] and *Why are there fewer board games with a triangular grid?* [54].

The exact calculations for the percentage of a regular polygon closer to its center than its perimeter can be found at *What is the probability that a point chosen randomly from inside an equilateral triangle is closer to the center than to any of the edges?* [53].

A discussion of the drunken walk phenomena can be found at *The Problem with Hexagons* [50].

Polyominoes: The book *Polyominoes: Puzzles, Patterns, Problems, and Packings - Revised and Expanded Second Edition* [30] is a broad introduction to the subject of polyominoes.

Graphs The posting *Design Diary: Undaunted: Stalingrad, or A Nondestructive Game of Epic Conflict* [22] on `boardgamegeek` describes how graphs were used in developing the game *Undaunted: Stalingrad*.

Probability: There is a wealth of content related to probability and games. This is not surprising, as early probabilists frequently focused on gambling questions. The book *Uncertainty in Games* [20] is aimed at game designers and introduces how uncertainty is used in games. The papers "*Carcassonne* in the Classroom" [17] and "Classroom and Computational

Investigations of *Camel Up*" [19] discuss how to use these tabletop games to teach probability in the classroom.

Markov processes: Markov processes are frequently used to analyze the outcome of multi-round battles. There are analyses for *Risk* in [43, 44, 48] and for *War of the Ring* (p.75) at [3]. The website from which that article comes, [2], contains several posts that explore probability as it appears in modern tabletop games.

In the text, I mentioned that Markov processes can be used to analyze the probability of reaching spaces in a roll-and-move game, you can find analyses for *Snakes and Ladders* in [4], *Candy Land* in [41], *Chutes and Ladders* in [18, 27], and *Monopoly* in [1, 6].

Game Theory: An introduction to the proof that *Hex* always results in a win for the first player in optimal play can be found in "The Game of Hex and the Brouwer Fixed-Point Theorem" [28].

The analysis of threats and promises is taken from "A Game-Theoretic Rendering of Promises and Threats" [38].

The *Prisoner's Dilemma* is not restricted to two-player games. See "Perils of the Football Draft" [23] for an example of a three-player version. A discussion of the iterated prisoner's dilemma (where two players play the game repeatedly) can be found in *The Evolution of Cooperation* [7].

"The Games Game Theorists Play" [13] discusses a very pure game of alliances and negotiation known as *So Long Sucker*, which was introduced in the paper "So Long Sucker—A Four Person Game" [33].

Dollar Auction: The unexpected results of the dollar auction can be found in "The Dollar Auction Game: A Paradox in Noncooperative Behavior and Escalation" [46].

Gnome Hat Games: A surprising amount of mathematical literature concerns gnomes who cannot see the hats on their heads, for example, [9, 11, 15, 29, 55] all explore this topic.

Bibliography

[1]   Stephen D. Abbott and Matt Richey. "Take a Walk on the Boardwalk." In: *The College Math Journal* 28.3 (May 1997), pp. 162–171.

[2]   Tim Adamson. *Quantifying Strategy: The Mathematics of Tabletop Games*. 2020. URL: https://www.quantifyingstrategy.com/.

[3]   Tim Adamson. *How Long will a Siege Last in War of the Ring?* 2022. URL: https://www.quantifyingstrategy.com/2022/05/how-long-will-siege-last-in-war-of-ring.html.

[4]   S. C. Althoen, L. King, and K. Schilling. "How Long Is a Game of Snakes and Ladders?" In: *The Mathematical Gazette* 77.478 (Mar. 1993), pp. 71–76.

[5]   M. A. Armstrong. *Groups and Symmetry*. Undergraduate Texts in Mathematics. Springer Science+Business Media, LLC, 1988.

[6]   Robert B. Ash and Richard L. Bishop. "Monopoly as a Markov Process." In: *Mathematics Magazine* 45.1 (Jan. 1972), pp. 26–29.

[7]   Robert Axelrod. *The Evolution of Cooperation*. Basic Books, Inc., 1984.

[8]   John D. Beasley. *The Mathematics of Games*. Dover Publications Inc., 2006.

[9]   M. Bernstein. "The Hat Problem and Hamming Codes." In: *FOCUS* 21 (2001), p. 4.

[10] Jörg Bewersdorff. *Luck, Logic, and White Lies*. CRC Press, 2022.

[11] Ezra Brown and James Tanton. "A Dozen Hat Problems." In: *Math Horizons* 16.4 (2009), pp. 22–25.

[12] Richard A. Brualdi. *Introductory Combinatorics*. 5th ed. Pearson, 2009.

[13] D. Graham Burnett. "The Games Game Theorists Play." In: *Cabinet* (2012).

[14] William Burnside. *Theory of Groups of Finite Order*. Cambridge: At the University Press, 1897.

[15] Steve Butler et al. "Hat Guessing Games." In: *SIAM J. of Discrete Math* 22 (2008), p. 592.

[16] Mindy Capaldi, ed. *Teaching Mathematics Through Games*. MAA Press, 2021.

[17] Mindy Capaldi and Tiffany Kolba. "*Carcassonne* in the Classroom." In: *The College Math Journal* 48.4 (Sept. 2017), pp. 265–273.

[18] Leslie A. Cheteyan, Stewart Hengeveld, and Michael A. Jones. "Chutes and Ladders for the Impatient." In: *The College Math Journal* 42.1 (Jan. 2011), pp. 2–8.

[19] Thomas J. Clark. "Classroom and Computional Investigations of *Camel Up*." In: *The College Math Journal* 52.4 (2021), pp. 289–296.

[20] Greg Costikyan. *Uncertainty in Games*. The MIT Press, 2013. ISBN: 0262018969.

[21] Christopher Cox et al. "How to Make the Perfect Fireworks Display: Two Strategies for Hanabi." In: *Mathematics Magazine* 88.5 (2015), pp. 323–336. ISSN: 0025570X, 19300980.

[22] *Design Diary: Undaunted: Stalingrad, or A Non-destructive Game of Epic Conflict*. 2022. URL: https://boardgamegeek.com/blogpost/138745/design-diary-undaunted-stalingrad-or-non-destructi.

[23] Dr. Crypton. "Perils of the Football Draft." In: *Science Digest* (July 1986), pp. 76–79.

[24] David Easley and Jon Kleinberg. *Networks, Crowds, and Markets: Reasoning about a Highly Connected World*. Cambridge University Press, 2010.

[25] Geoff Engelstein. *Gametek: The Math and Science of Gaming*. Ludology, 2017.

[26] Geoff Engelstein. *Achievement Relocked: Loss Aversion and Game Design*. The MIT Press, 2020.

[27] Steve Gadbois. "Mr. Markov plays Chutes and Ladders." In: *The UMAP Journal of Undergraduate Mathematics and Its Applications* 14.1 (1993), pp. 31–38.

[28] David Gale. "The Game of Hex and the Brouwer Fixed-Point Theorem." In: *The American Mathematical Monthly* 86.1 (Dec. 1979), pp. 818–827.

[29] Martin Gardner. *Penrose Tiles to Trapdoor Ciphers . . . and the Return of Dr. Matrix*. Recreational Mathematics. New York: W. H. Freeman and Company, 1989, pp. x+311.

[30] Solomon W. Golomb. *Polyominoes: Puzzles, Patterns, Problems, and Packings - Revised and Expanded Second Edition*. NED - New edition. Vol. 111. Princeton University Press, 1994. ISBN: 9780691085739.

[31] Branko Grünbaum and Geoffrey C. Shephard. "Tilings by Regular Polygons." In: *Mathematics Magazine* 50.5 (Nov. 1977), pp. 227–247.

[32] John Harris, Jeffry L. Hirst, and Michael Mossinghoff. *Combinatorics and Graph Theory*. Undergraduate Texts in Mathematics. Springer Science+Business Media, LLC, 2008.

[33] Mel Hausner et al. "So Long Sucker—A Four Person Game." In: *Game Theory and Related Approachs to Social Behavior: Selections*. Ed. by Martin Shubik. Wiley, 1964, pp. 359–361.

[34] *Hyperbolica*. 2022. URL: https://steamcommunity.com/app/1256230.

[35] Dan Jolin. "The rise and rise of tabletop gaming." In: *The Guardian* (Sept. 25, 2016). URL: https://www.theguardian.com/technology/2016/sep/25/board-games-back-tabletop-gaming-boom-pandemic-flash-point.

[36] Jonathan Kay. "The Invasion of the German Board Games." In: *The Atlantic* (Jan. 21, 2018). URL: https://www.theatlantic.com/business/archive/2018/01/german-board-games-catan/550826/.

[37] Jon-Lark Kim. *Mathematicians Playing Games*. 1st ed. A K Peters/CRC Press, 2023.

[38] Daniel B. Klein and Brendan O'Flaherty. "A Game-Theoretic Rendering of Promises and Threats." In: *Journal of Economic Behavior & Organization* 21.3 (1993), pp. 295–314. ISSN: 0167-2681.

[39] Eric Lehman, F. Thomson Leighton, and Albert R. Meyer. *Mathematics for Computer Science*. MIT Open Courseware, 2015.

[40] S. J. Macdonald. *Board Game BBQ*. Podcast. Mar. 2023. URL: https://boardgamebbq.com/234-special-guest-sam-macdonald/.

[41] Kamar Mack et al. "Mr. Markov Tours Candy Land." In: *The UMAP Journal of Undergraduate Mathematics and Its Applications* 35.1 (2014), pp. 11–21.

[42] Ben Orlin. *Math Games with Bad Drawings*. Black Dog & Leventhal Publishers, 2022.

[43] Jason A. Osborne. "Markov Chains for the RISK Board Game Revisited." In: *Mathematics Magazine* 76.2 (Apr. 2003), pp. 129–135.

[44] Pamela Pierce and Robert Wooster. "Conquer the World with Markov Chains." In: *Math Horizons* 22.4 (Apr. 2015), pp. 18–21.

[45] Frederick Reiber. "Major Developments in the Evolution of Tabletop Game Design." In: *2021 IEEE Conference on Games (CoG)*. 2021, pp. 1–8. DOI: 10.1109/CoG52621.2021.9619158.

[46] Martin Shubik. "The Dollar Auction Game: A Paradox in Noncooperative Behavior and Escalation." In: *The Journal of Conflict Resolution* 15.1 (Mar. 1971), pp. 109–111.

[47] Steven Tadelis. *Game Theory: An Introduction*. Princeton University Press, 2013.

[48] Barış Tan. "Markov Chains and the RISK Board Game." In: *Mathematics Magazine* 70.5 (Dec. 1997), pp. 349–357.

[49] David G. Taylor. *Games, Gambling, and Probability: An Introduction to Mathematics.* CRC Press, 2021.

[50] *The Problem with Hexagons.* 2018. URL: https://www.general-staff.com/the-problem-with-hexagons/.

[51] David Thompson. *Email.* personal communication. Nov. 2022.

[52] *Triangular Grids - there are almost no games that use it. Only hex and square grids. Have you heard of any games using triangles, do you have experience of playing/developing these?* 2020. URL: https://www.reddit.com/r/gamedev/comments/hokdvq/triangular_grids_there_are_almost_no_games_that/.

[53] *What is the probability that a point chosen randomly from inside an equilateral triangle is closer to the center than to any of the edges?* 2016. URL: https://math.stackexchange.com/questions/1688936/what-is-the-probability-that-a-point-chosen-randomly-from-inside-an-equilateral/1706984#1706984.

[54] *Why are there fewer board games with a triangular grid?* 2022. URL: https://boardgames.stackexchange.com/questions/633/why-are-there-fewer-board-games-with-a-triangular-grid.

[55] Peter Winkler. *Mathematical Puzzles: A Connoisseur's Collection.* A K Peters/CRC Press, 2003.

[56] Stewart Woods. *Eurogames: The Design, Culture and Play of Modern Board Games.* McFarland & Company, Inc., 2012.

Referenced Games

Games listed may be copyrighted, registered, or trademarked by their respective designers and/or publishers. The publishers listed here refer to the publisher responsible for the edition of the game in my collection. Information about other editions may be found at `BoardGameGeek`.

Several very good games are *not* listed here, including many of my favorites. There isn't enough room to cover all the fantastic games available now. You can explore websites, podcasts, and videos to find more great games. Feel free to suggest any games you wish me to look at using my email: `magister_ludi@monsterworks.com`.

Architects of the West Kingdom. S. J. Macdonald and Shem Phillips. *Garphill Games*, 2018. *2018 Golden Geek Board Game of the Year Nominee, 2019 Kennerspiel des Jahres Recommended* (p. 174)

Ark Nova. Mathias Wigge. *Capstone Games & Feuerland Spiele*, 2021. *2021 Golden Geek Heavy Game of the Year Winner, 2022 Deutscher Spiele Preis Best Family/Adult Game Winner, 2022 Japan Boardgame Prize Voters' Selection Winner, 2022 Kennerspiel des Jahres Recommended* (pp. xviii, 11, 12, 21, 38, 39)

Arkham Horror: The Card Game. Maxine Juniper Newman and Nate French. *Fantasy Flight Games*, 2016. *2016 Golden Geek Best Card Game Winner, 2017 International Gamers Award - General Strategy: Two-players Winner, 2017 SXSW Tabletop Game of the Year Winner, 2017 RPC Fantasy Award Board Game Winner* (p. xv)

Azul. Michael Kiesling. *Next Move Games*, 2017. *2018 Spiel des Jahres Winner* (pp. 119–123)

Beast. Aron Midhall, Elon Midhall, and Assar Pettersson. *Studio Midhall*, 2023

Black Orchestra. Philip duBarry. *Game Salute*, 2016. *2016 Golden Geek Best Cooperative Game Nominee* (pp. 99–101)

Bohnanza. Uwe Rosenberg. *AMIGO*, 1997. *1997 Spiel des Jahres Recommended* (pp. 10, 13)

Brass: Birmingham. Gavan Brown, Matt Tolman, and Martin Wallace. *Roxley Games*, 2018. *2018 Golden Geek Best Strategy Board Game Winner* (pp. xvii, 1, 2, 18, 21, 72)

Candy Land. Eleanor Abbott. *Hasbro*, 1949 (pp. xi, 22, 92, 93, 114, 175, 176)

Carcassonne. Klaus-Jürgen Wrede. *Hans im Glück*, 2000. *2001 Deutscher Spiele Preis Best Family/Adult Game Winner, 2001 Spiel des Jahres Winner* (pp. xi, xviii, 46, 53–55, 57, 59, 175)

Cascadia. Randy Flynn. *Flatout Games & Alderac Entertainment Group*, 2021. *2021 Golden Geek Light Game of the Year Winner, 2022 Geek Media Awards Family Game of the Year Winner, 2022 Spiel des Jahres Winner* (pp. xvii, 7–10, 13, 21, 55, 56)

Catan. Klaus Teuber. *CATAN GmbH*, 1995. *1995 Spiel des Jahres Winner, 1996 Origins Awards Best Fantasy or Science Fiction Board Game Winner, 2001 Origins Awards Hall of Fame Inductee* (pp. xi, xvii, 34, 175)

Checkers. 1150 (pp. xi, 118)

Chess. 1475 (pp. xi, 22, 32, 117, 119)

Chicken. (undated) (pp. 117, 133–136)

Chutes and Ladders. *Milton Bradley*, 1943 (pp. xi, 114, 176)

Clue. Anthony E. Pratt. *Hasbro*, 1949. *2011 Kinderspielexperten "8-to-13-year-olds" Winner* (pp. xi, 151)

Codenames. Vlaada Chvátil. *Czech Games Edition*, 2015. *2015 Golden Geek Best Family Board Game Winner, 2015 Golden Geek Best Party Board Game Winner, 2016 Spiel des Jahres Winner* (pp. 158, 162, 163, 170)

Cryptid. Hal Duncan and Ruth Veevers. *Osprey Games*, 2018. *2018 Golden Geek Most Innovative Board Game Nominee, 2022 Kennerspiel des Jahres Nominee* (pp. 151, 152)

Dominoes. 1500 (pp. xi, 28)

Empire of the Sun. Mark Herman. *GMT Games*, 2005. *2005 Walter Luc Haas Best Simulation Winner, 2005 Charles S. Roberts Best World War II Boardgame Winner* (p. xii)

Flamme Rouge. Asger Harding Granerud. *Lautapelit.fi*, 2016. *2017 Guldbrikken Best Adult Game Winner, 2018 Gioco dell'Anno Winner* (pp. 13, 15–17, 21, 101, 102)

For Sale. Stefan Dorra. *2 Pionki*, 1997. *2017 Hungarian Board Game Award Nominee, 2016 Lys Grand Public Finalist* (pp. 146, 147)

Fury of Dracula. Frank Brooks, Stephen Hand, and Kevin Wilson. *Fantasy Flight Games*, 2015. *2015 Golden Geek Best Thematic Board Game Nominee* (pp. 63–67, 70, 75)

Galaxy Trucker. Vlaada Chvátil. *Czech Games Edition*, 2007. *2008 Spiel des Jahres Recommended* (pp. xviii, 45, 46, 57, 58)

Gloomhaven. Isaac Childres. *Cephalofair Games*, 2017. *2017 Golden Geek Board Game of the Year Winner, 2018 Origins Awards Game of the Year Winner, 2018 SXSW Tabletop Game of the Year Winner* (pp. 33, 84, 85)

Go. 2200 BCE (pp. xi, 22, 34, 117, 119)

Hamlet: The Village Building Game. David Chircop. *Mighty Boards Ltd,* 2022 (pp. xvii, 38, 39)

Hanabi. Antoine Bauza. *R & R Games,* 2010. *2013 Spiel des Jahres Winner,* (pp. xviii, 158, 161, 164, 172)

Hex. Piet Hein and John Nash, 1942 (pp. 118, 119, 174, 176)

High Frontier 4 All. Phil Ekland et al. *Ion Game Design & Sierra Madre Games,* 2020. *2020 Golden Geek Heavy Game of the Year Nominee* (p. xii)

Hoplomachus: Remastered. Adam Carlson, Josh J. Carlson, and Logan Giannini. *Chip Theory Games,* 2023 (pp. xvii, 32, 33, 63, 85, 94)

King of Tokyo. Richard Garfield. *IELLO USA LLC,* 2011. *2012 Golden Geek Best Children's Game Winner, 2012 Golden Geek Best Family Board Game Winner, 2012 Golden Geek Best Party Game Winner, 2013 Guldbrikken Best Family Game Winner, 2013 Nederlandse Spellenprijs Best Family Game Winner, 2014 Gra Roku Game of the Year Winner* (pp. 97, 99)

Letter Jam. Ondra Skoupý. *Czech Games Edition,* 2019. *2019 Golden Geek Most Innovative Board Game Nominee, 2019 Golden Geek Best Party Game Nominee, 2019 Golden Geek Best Cooperative Game Nominee* (pp. 158, 169, 170, 172)

Mage Knight. Vlaada Chvátil. *WizKids,* 2011. 2012 Golden Geek Best Thematic Board Game Winner, 2012 Golden Geek Best Strategy Board Game Nominee (p. xv)

Modern Art. Reiner Knizia. *Hans im Glück,* 1992. *1993 Deutscher Spiele Preis Best Family/Adult Game Winner, 1993 Spiel des Jahres Recommended,* (pp. 138, 140, 141)

Monopoly. Charles Darrow and Elizabeth J. Magie (Philips). *Parker Brothers,* 1935 (pp. xi, 92, 93, 114, 176)

My City. Reiner Knizia. *KOSMOS,* 2020. *2020 Spiel des Jahres Nominee, 2020 Golden Geek Most Innovative Board Game Nominee, 2020 Golden Geek Medium Game of the Year Nominee,* (pp. xvii, 37, 38)

Nemesis. Adam Kwapiński. *Awaken Realms,* 2018. *2019 Board Game Quest Awards Best Thematic Game Winner* (pp. xviii, 116, 135)

Nidavellir. Serge Laget. *GRRRE Games,* 2020. *2020 Tric Trac d'Or, 2020 Swiss Gamers Award Winner* (p. 146)

Oathsworn: Into the Deep Wood. Jamie Jolly. *Shadowborne Games LLC,* 2022. *2022 Golden Geek Best Thematic Board Game Nominee, 2022 Golden Geek Best Cooperative Game Nominee, 2023 UK Games Expo Best Boardgame (American-Style) Winner* (pp. xviii, 33, 86, 87, 94, 103–106)

Pandemic. Matt Leacock. *Z-Man Games, Inc.* 2008. *2009 Golden Geek Best Family Board Game Winner, 2009 Spiel des Jahres Nominee, 2011 MinD-Spielepreis Winner* (p. 83)

Pandemic Legacy: Season 1. Rob Daviau and Matt Leacock. *Z-Man Games, Inc.* 2015. *2015 Golden Geek Best Strategy Board Game Winner, 2015 Golden Geek Best Thematic Board Game Winner, 2015 Golden Geek Board Game of the Year Winner, 2016 Kennerspiel des Jahres Nominee, 2016 SXSW Tabletop Game of the Year Winner* (pp. 83, 84)

Prisoner's Dilemma. (undated) (pp. 117, 133–136, 176)

Project L. Michal Mikeš, Jan Soukal, and Adam Španěl. *Boardcubator, sro,* 2020. *2022 MinD-Spielepreis Short Game Winner* (pp. xviii, 78–81)

QE. Gavin Birnbaum. *BoardGameTables,* 2019. *2023 Spiel des Jahres Recommended* (pp. 140, 142)

Ra. Reiner Knizia. *25th Century Games, LLC,* 2022. *1999 Meeples' Choice Award, 2016 Juego del Año Recommended* (pp. xviii, 143–145)

Railroad Ink. Hjalmar Hach and Lorenzo Silva. *Horrible Games,* 2018. *2019 5 Seasons Best International Roll & Write Game Winner* (pp. xviii, 51, 54)

Rising Sun. Eric M. Lang. *CMON Productions Limited,* 2018. *2018 RPC Fantasy Award Tabletop & Miniatures Game Winner, 2018 UK Games Expo Best Board Game (American Style) People's Choice Winner, 2018 UK Games Expo Best Board Game (American Style) Judges Award Winner* (pp. 125–128, 146)

Risk. Albert Lamorisse and Michael I. Levin. *Hasbro,* 1959 (pp. xi, 176)

Root. Cole Wehrle. *Leder Games,* 2018. *2018 Golden Geek Best Thematic Board Game Winner, 2018 Golden Geek Board Game of the Year Winner, 2018 Golden Geek Most Innovative Board Game Winner, 2019 Juego del Año Recommended, 2019 SXSW Tabletop Game of the Year Winner, 2019 UK Games Expo Best Board Game (Strategic Style) Judges Award Winner* (pp. 130–132)

SCOUT. Kei Kajino. *Oink Games,* 2019. *2022 Spiel des Jahres Nominee* (pp. 10, 13, 21, 80, 82, 83)

Scythe. Jamey Stegmaier. *Stonemaier Games,* 2016. *2016 Golden Geek Board Game of the Year Winner, 2016 Golden Geek Best Strategy Board Game Winner, 2016 Golden Geek Best Thematic Board Game Winner* (pp. 123–125, 146)

Snakes and Ladders. (undated) (pp. 114, 176)

So Long Sucker. Mel Hausner et al., 1950 (pp. 148, 176)

Star Wars: Rebellion. Corey Konieczka. *Fantasy Flight Games,* 2016. *2016 Golden Geek Best 2-Player Board Game Winner, 2016 UK Games Expo Best Boardgame with Miniatures Winner* (pp. 107, 108, 111–114)

Tammany Hall. Doug Eckhart. *StrataMax Games & Pandasaurus Games,* 2007 (pp. xviii, 137, 138, 145, 146)

The Grizzled. Fabien Riffaud and Juan Rodriguez. *Sweet November,* 2015. *2017 Kennerspiel des Jahred Recommended* (pp. 2, 3)

The Isle of Cats. Frank West. *The City of Games,* 2019. *2019 Golden Geek Board Game of the Year Nominee, 2019 Golden Geek Best Family Board Game Nominee* (pp. xvii, 38)

The Search for Planet X. Ben Rosset and Matthew O'Malley. *Foxtrot Games LLC.* 2022. *2021 SXSW Tabletop Game of the Year Nominee, 2021 American Tabletop Strategy Games Recommended* (pp. xviii, 151, 154–157)

The Shipwreck Arcana. Kevin Bishop. *Meromorph Games,* 2017 (pp. xviii, 150, 151, 158–160)

Through the Ages: A New Story of Civilization. Vlaada Chvátil. *Czech Games Edition,* 2015. *2015 Golden Geek Board Game of the Year Nominee* (pp. 138, 142)

Tic Tac Toe. 1800 BCE (pp. xi, 118, 119)

Ticket To Ride. Alan R. Moon. *Days of Wonder,* 2004. *2004 Origins Awards Best Board Game Winner, 2004 Spiel des Jahres Winner* (pp. xi, xviii, 62, 63, 71–75, 77, 78, 175)

Tri-Ominos. Allan Cowan. *Pressman Toy Corp.* 1965 (p. 28)

Turing Machine. Fabien Gridel and Yoann Levet. *Le Scorpion masqué inc,* 2022. *2022 Golden Geek Best Solo Board Game Winner, 2023 American Tabletop Casual Games Winner* (pp. 151–154)

Twilight Struggle. Ananda Gupta and Jason Matthews. *GMT Games,* 2005. *2006 Golden Geek Best 2-Player Board Game Winner, 2006 Golden Geek Best Wargame Winner* (p. 92)

Undaunted: Battle of Britain. David Thompson and Trevor Benjamin. *Osprey Publishing, Ltd.* 2022 (pp. 24, 33)

Undaunted: Normandy. Trevor Benjamin and David Thompson. *Osprey Publishing, Ltd.* 2019. *2019 Golden Geek Best Wargame Winner* (pp. xvii, 24, 25, 36, 63)

Undaunted: Stalingrad. David Thompson and Trevor Benjamin. *Osprey Publishing, Ltd.* 2021. *2022 Golden Geek Best Wargame Winner* (pp. xv, 24, 175)

War. (undated) (p. 92)

War of the Ring. Roberto Di Meglio, Marco Maggi, and Francesco Nepitello. *Ares Games Srl,* 2011. *2007 Golden Geek Best Wargame Nominee, 2006 International Gamers Award - General Strategy: Two-players* (pp. 75, 94, 176)

Watergate. Matthias Cramer. *Capstone Games & Frosted Games,* 2019. *2019 Golden Geek Best 2-Player Board Game Winner* (pp. xvii, 33, 34)

Yahtzee. Edwin S. Lowe. *Milton Bradley,* 1956 (pp. xi, 92, 98)

Zombicide: Black Plague. Raphaël Guiton, Jean-Baptiste Lullien, and Nicolas Raoult. *CMON Global Limited & Guillotine Games,* 2015. *2015 Golden Geek Best Solo Board Game Nominee* (pp. 87, 88, 90–97, 104)

Index

Printed in the United States
by Baker & Taylor Publisher Services